BetterCloud

Controlling Your SaaS Environment

A Six-Part Framework for Effectively Managing and Securing SaaS Applications

FIRST EDITION

by David Politis

Controlling Your SaaS Environment: A Six-Part Framework for Effectively Managing and Securing SaaS Applications, First Edition is published by BetterCloud.

Author: David Politis | Editor: Christina Wang

www.bettercloud.com
www.controlyoursaas.com

TABLE OF CONTENTS

EXECUTIVE SUMMARY

Software-as-a-Service (SaaS) adoption is on the rise. Companies use 16 SaaS applications on average today, up 33% from 2016. In fact, 38% of organizations say they're already running nearly all (80% or more) of their business applications on SaaS, and that figure is estimated to rise to 73% by 2020. But while SaaS significantly boosts productivity and collaboration, it also brings unprecedented challenges that IT teams are encountering unilaterally. As a result, the shift to SaaS is driving a need for new frameworks and tools to manage, secure, and support mission-critical applications.

Controlling Your SaaS Environment was created by synthesizing insights from interviews, surveys, and conversations with thousands of IT professionals over the last three years. The book introduces the SaaS Application Management and Security Framework™, the first framework of its kind, which proposes novel, innovative solutions to several key challenges that IT professionals are facing in SaaS environments.

The world is moving to SaaS, whether we like it or not. But this shift brings about a completely new paradigm for IT teams, and *Controlling Your SaaS Environment* is the first text to fully outline how IT must fundamentally rethink how they approach management and security in modern workplaces. ▲

By reading *Controlling Your SaaS Environment*, you will:

- ▲ Understand the key challenges faced by modern IT professionals, and why the status quo for SaaS application management and security cannot continue
- ▲ Learn best practices for IT on managing, securing, and supporting modern workplace technology
- ▲ Envision what an "enlightened" state for IT would look like when the framework is fully executed
- ▲ Connect the framework to real-life news stories and data that illustrate the potential risks and costs of failing to manage and secure SaaS applications more holistically
- ▲ Visualize how all six elements of the framework build upon each other to create a high-performing IT environment
- ▲ Explore the range of solutions for executing the framework, including different vendors' advantages, disadvantages, and functionalities

INTRODUCTION

The Data Breach That Prompted a US Politician to Say: "It Appears These 'Experts' Need to Learn a Thing or Two About Protecting Sensitive Information"

On a brisk fall day in October 2015, an administrator at a midsize company enabled an option in Slack, a real-time messaging application. The option was innocuous; all it did was allow Slack to automatically generate document previews when employees shared Google Drive documents and items in Slack. This was extremely common; many companies enabled this option for their Slack and Google Drive instances.

In March 2016—five months later—a supervisor discovered that this option was active. By then, 100 employees had enabled this connection. And by then, senior security officers were alarmed.

Why?

Unbeknownst to the administrator, sharing links to Google Drive files in Slack also put those documents on Slack's databases. In order for the previews to be created and made searchable, Slack automatically indexed and stored them for easy searching and reference purposes.

By integrating the two applications with a single click, the administrator had exposed 100 Drive accounts and caused a potentially disastrous data breach.

By integrating the two applications with a single click, the administrator had exposed **100 Drive accounts and caused a potentially disastrous data breach**.

To make matters worse, this midsize organization was 18F, a prominent digital services team that is part of the General Services Administration (GSA). As part of the federal government, they needed to exercise extra caution when connecting applications and securing sensitive data.

The inspector general's office swiftly released a report about the data breach, stating that "Over one hundred GSA Google Drives were reportedly accessible by users both inside and outside of GSA during a five month period, potentially exposing sensitive content such as personally identifiable information and contractor proprietary information."[1] Almost a year later, a follow-up report read: "As a result of our alert report, GSA has since confirmed that content containing personally identifiable information (PII) was exposed to unauthorized users as a result of this breach."[2]

Jason Chaffetz (R-UT), House Oversight and Government Reform Committee Chairman at the time, called the incident "alarming." He released a statement lambasting the administrators, saying: "While we appreciate the efforts to recruit IT talent into the federal government, it appears these 'experts' need to learn a thing or two about protecting sensitive information."[3]

But in this era of Software-as-a-Service (SaaS), "learning a thing or two about protecting sensitive information" is not as easy as it sounds. The administrators should have been able to prevent this from happening, but they couldn't. The problem wasn't that they were inordinately lax or inept. The problem is that IT is encountering new, unprecedented challenges with SaaS, and IT professionals lack the appropriate tools to solve those challenges.

Likewise, they also lack knowledge. They don't know what operational security entails; how data spreads, overlaps, interacts; where it lives and where it's exposed; how applications operate with each other. They do not have a framework that explains how to think about their SaaS environments.

---

1 Office of Inspections and Forensic Auditing and Office of Inspector General. "Management Alert Report: GSA Data Breach." US General Services Administration, https://www.gsaig.gov/sites/default/files/ipa-reports/Alert Report-GSA Data Breach 5.12.16.pdf. Accessed 1 July 2017.
2 Office of Inspections and Forensic Auditing and Office of Inspector General. "Evaluation of 18F's Information Technology Security Compliance." https://www.gsaig.gov/sites/default/files/ipa-reports/OIG%20EVALUATION%20REPORT_Evaluation%20of%2018F%20IT%20Security%20Compliance_JEF17-002_February%2021%202017.pdf. Accessed 1 July 2017.
3 "Press Release: Chaffetz Statement on GSA 18F Flash Alert." Committee on Oversight and Government Reform, https://oversight.house.gov/release/chaffetz-statement-on-gsa-18f-flash-alert/. Accessed 30 June 2017.

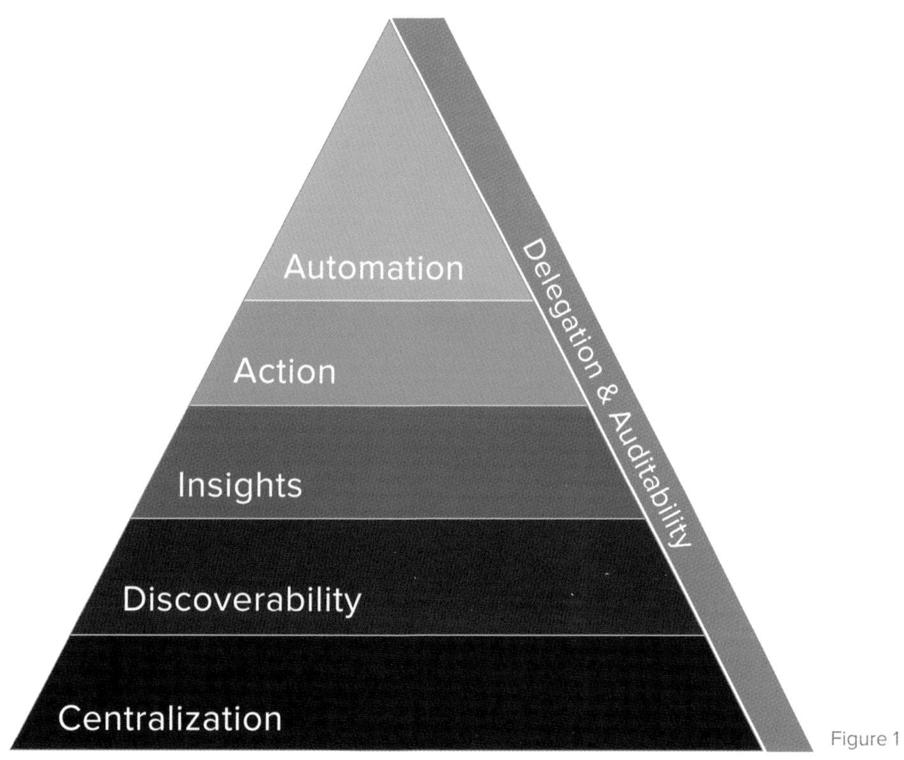

Figure 1

SaaS Application Management and Security Framework

A Framework for Thinking About the Runaway Train That Is SaaS

This story about 18F does not describe an isolated incident. It is happening everywhere.

As such, IT needs to get a firm grasp on the ins and outs of SaaS management and operational security. SaaS is a runaway train, a pure juggernaut in the cloud computing market. The largest public cloud market in 2017 will be SaaS, which is forecasted to reach $75.7 billion by 2020.[4] SaaS adoption is on the rise—companies use 16 SaaS applications on average today, up 33% from 2016[5]—and isn't slowing anytime soon. For almost every

4 "Gartner Says Worldwide Public Cloud Services Market to Grow 18 Percent in 2017." Gartner, http://www.gartner.com/newsroom/id/3616417. Accessed 15 June 2017.
5 "2017 State of the SaaS-Powered Workplace Report." BetterCloud, https://www.bettercloud.com/monitor/state-of-the-saas-powered-workplace-report/. Accessed 15 June 2017.

business unit or use case, there now exists a SaaS solution or vendor. While SaaS increases productivity, its rapid adoption also makes IT environments increasingly fragmented and unwieldy. As a result, the SaaS explosion is changing how IT professionals must think about SaaS going forward.

What's missing currently is a structured, comprehensive way to think about managing enterprise SaaS software in order to keep up with a **rapidly evolving workplace**.

But what should that shift in thinking entail? That's a bit murkier. In the absence of established, demonstrable best practices around SaaS management and operational security, IT is entering uncharted territory.

What's missing currently is a structured, comprehensive way to think about managing enterprise SaaS software in order to keep up with a rapidly evolving workplace. There is no framework that outlines a modern-day approach to SaaS management nor are there guiding principles that help IT professionals achieve their security goals and requirements in SaaS environments. As environments shift from being homogeneous to heterogeneous, IT's management and operational strategies must also shift accordingly. The shift to SaaS is driving a need for new frameworks and tools to manage, secure, and support mission-critical applications.

This framework—the first of its kind—was created by synthesizing insights from interviews, surveys, and conversations with thousands of IT professionals over the last three years. These IT practitioners have varying tenure and hail from a wide range of industries, geographical locations, and organization sizes, but they have one quality in common: all are directly involved in core IT management and operational security.

What emerged was a clear, consistent portrait. These IT practitioners all spoke of nascent challenges around SaaS management and security that they are encountering unilaterally as their organizations adopt SaaS applications. The SaaS Application Management and

Security Framework will address those challenges at length, as well as the strategic elements IT needs to solve them. It will also describe the costs and risks of operating outside the framework. More broadly, it teaches IT professionals how to think about the way data interacts, whether it's in a single SaaS, multi-instance, or multi-SaaS environment. These concepts are equally relevant to IT professionals who manage one SaaS application or dozens.

Visualized in Figure 1, the framework illustrates how IT professionals should think about engaging with their SaaS applications. It operates on the assumption that IT is treating SaaS platforms as the systems of record for its company's data, either in one or two areas of the business or across the entire business. Each element builds upon the previous one and each is key to managing, securing, and supporting a SaaS environment. Collectively, the elements in this framework form a playbook on operational security and management in the SaaS era. ▲

Note: This framework addresses the day-to-day management and operational security challenges for SaaS applications that are already deployed. It will not cover how to select and/or deploy SaaS applications.

CHAPTER ONE

CENTRALIZATION

CHAPTER HIGHLIGHTS

CENTRALIZATION

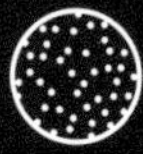

SaaS applications are creating a massive information sprawl the likes of which IT has never seen before.

SaaS sprawl leads to a **loss of administrative control**.

IT and security teams no longer natively have root access to users, data, files, and settings. That access is available purely via APIs or not at all. This inhibits their ability to even see what's in their SaaS environment, let alone troubleshoot or fully secure it.

IT teams cannot control an environment they do not have visibility into, nor can they take any meaningful action unless they see all the data centralized in one place first.

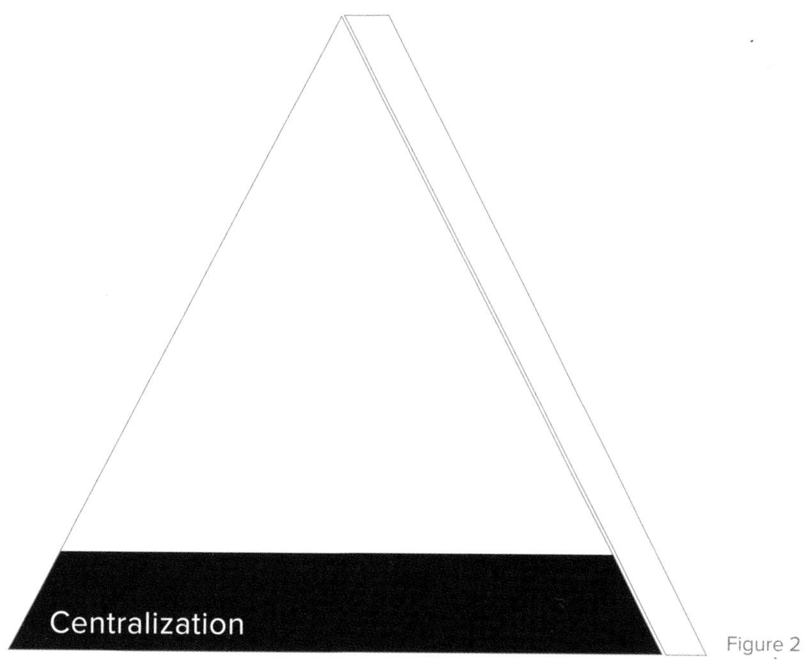

Figure 2

SaaS Application Management and Security Framework

CENTRALIZATION

In the legacy world, security was done from the inside out. IT secured everything inside the environment first—e.g., the computer someone was using, the servers that applications were running on, the internal network—and then moved on to the external network.

Today, the direction is reversed. IT must take an "outside-in" approach. IT must first secure the applications that are in the cloud and then configure all the data that lives outside of the network. Then IT can work on securing identity, external networks, and device management.

But managing and securing all the data that lives outside the network isn't easy. The first hurdle IT has to overcome is the sheer scale and complexity of the data sprawl.

Challenge #1: SaaS Applications Create a Complex, Interconnected Sprawl

SaaS is increasingly powering workplaces around the world. In fact, 38% of organizations say they're already running nearly all (80% or more) of their business applications on SaaS, and that figure is estimated to rise to 73% by 2020.[6] To many people, this is hardly surprising. SaaS applications are a major boon to productivity and collaboration, and they're easily interoperable from an end user perspective.

However, SaaS applications are also creating a massive information sprawl the likes of which IT has never seen before. As companies increasingly adopt SaaS applications as their systems of record, critical information is sprawled across a number of distinct sources rather than stored locally on physical endpoints such as users' devices and on-premises servers.

Figure 3

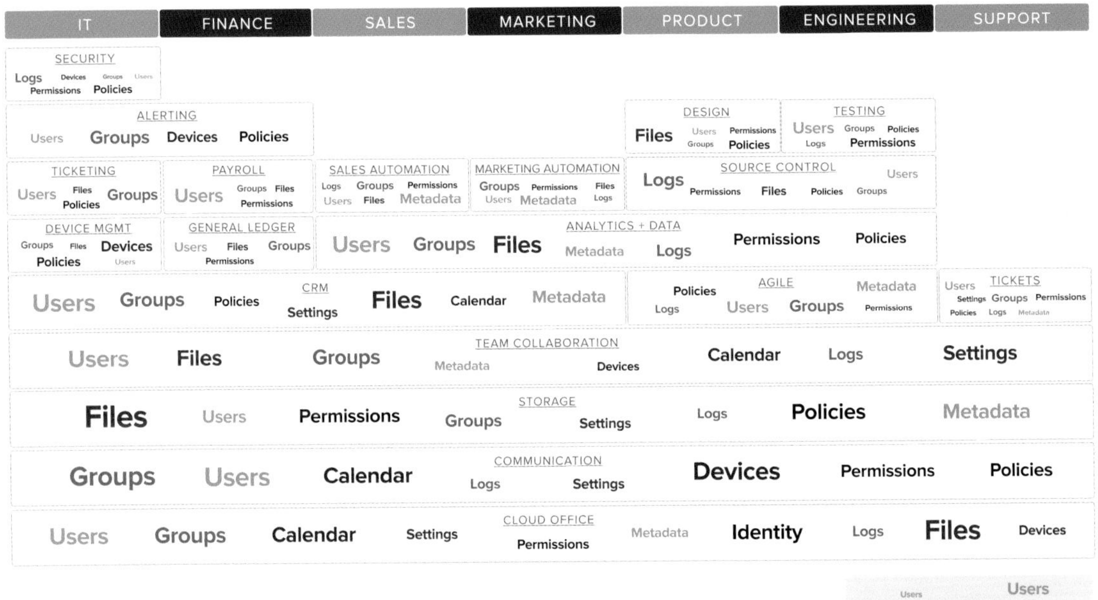

6 "2017 State of the SaaS-Powered Workplace Report." BetterCloud, https://www.bettercloud.com/monitor/state-of-the-saas-powered-workplace-report/. Accessed 15 June 2017.

Every SaaS application has a web of data objects that reference, interact, control, and/or rely on each other. Examples include users and groups, mailboxes, files, folders, records (e.g., tasks, opportunities, contacts, calendars), third-party applications (that have been installed from app marketplaces and authorized by users), logs, metadata, permissions, devices, policies, or phone numbers (e.g., numbers assigned to multiple endpoints like mobile applications, IP desk phones, and/or softphones that are included off an interactive voice response [IVR]/call tree).

Figure 3 visualizes what a typical SaaS stack looks like at a midsize organization. It depicts the types of applications often used within and across departments, as well as the types of data objects found in each application. The criticality of each data object is proportional to the size of the word (i.e., the larger the font size, the more important that data is). This figure illustrates the astronomical amount of data associated with SaaS applications, as well as how criticality varies across data objects.

This level of criticality depends on which type of application a data object belongs to. There are four general categories applications fall into:

- **Systems of record.** These are core business applications; they are the authoritative data source for a particular type of record, data object, or any given business process. Examples include CRM, HRIS, ERP, helpdesk, SSO applications, etc. Out of all four categories, this one is the most dense, as these core applications contain the most sensitive and critical data.

- **IT-sanctioned.** These applications are not necessarily systems of record, but are nevertheless sanctioned and managed by IT. Examples include business intelligence or videoconferencing software.

- **Connected.** These are applications that line of business users connect on their own and are not managed by IT. These applications are often geared toward productivity enhancements and as a result, the data objects here are less sensitive. However, they often connect directly to the system of record, accessing highly sensitive data without the same degree of IT oversight. IT departments do not typically have visibility into this data. Connected applications are often found and installed via vendor-supported marketplaces like the G Suite Marketplace, Salesforce AppExchange, Slack App Directory, and Zendesk Apps Marketplace.

- **Shadow IT.** Shadow IT consists of applications that are not explicitly approved by IT. These applications are installed by end users. For example, a user may create a HelloSign account to electronically sign a document, and then upload a contract with sensitive information into HelloSign without IT ever knowing about it. Other examples include messaging applications and browser add-ons.

Figure 3 is representative of the collective sprawl found in many SaaS environments. As an organization grows, the amount of SaaS application data will multiply by leaps and bounds. SaaS data is like the digital version of kudzu—its growth is rampant, it's notoriously hard to control, and it can have harmful consequences if left unchecked.

SaaS data is like the **digital version of kudzu**—its growth is rampant, it's notoriously hard to control, and it can have harmful consequences if left unchecked.

This web of data objects becomes inherently more complex in multi-SaaS environments. Data objects are connected (or overlap) across applications, but are not aware of each other, and there is no single place to view them all. As SaaS adoption grows, so does the amount of data living in SaaS applications, which in turn creates an enormous, decentralized information sprawl. Where SaaS data lives, and the questions of who has access to it and where it's exposed, become nebulous.

"The challenge with SaaS sprawl is that the necessary data is scattered across dozens of cloud apps, making it tough to find, analyze, and deploy. Think about it: Do you know exactly which cloud apps hold your customer data? Your employee data? Your supply chain data?" writes Informatica.[7] Shadow IT and connected applications make it nearly impossible to answer these questions.

This sprawl becomes even more amplified when companies have multiple instances of SaaS applications. For example, many companies have separate accounts of Slack, Zendesk, and/or Salesforce per department, business unit, or office location. The SaaS data in these environments is even more fragmented, as it is spread out and siloed across multiple instances of the same application.

"Know your data: if you don't know where your data resides, you can't protect it."
- CIO.com

IT professionals readily admit they lack visibility. A report by the Ponemon Institute, which surveyed more than 1,000 IT and IT security practitioners in the United States, United Kingdom, and Germany, found that 49% of respondents either did not agree or were unsure of whether they had visibility into employees' use of file sharing/file sync and share applications used in the workplace.[8]

Similarly, CIO.com writes, "Know your data: if you don't know where your data resides, you can't protect it."[9]

Unfortunately, IT can do neither. There is no central repository or configuration management database (CMDB) of all IT-sanctioned SaaS applications. There are many solutions such as cloud access security brokers (CASBs) and identity-as-a-service (IDaaS) offerings in the

7 Thompson, Graeme. "SaaS Sprawl: Is History Repeating Itself?" Informatica, https://blogs.informatica.com/2017/03/27/saas-sprawl-is-history-repeating-itself/. Accessed 30 June 2017.
8 "Breaking Bad: The Risk of Unsecure File Sharing." Ponemon Institute, https://www.intralinks.com/platform-solutions/solutions/via/breaking-bad-risk-unsecure-file-sharing. Accessed 7 July 2017.
9 Nichol, Peter B. "How CIOs prepare for tomorrow's healthcare data breaches." CIO, http://www.cio.com/article/3152861/security/how-cios-prepare-for-tomorrows-healthcare-data-breaches.html. Accessed 15 June 2017.

market that solve different parts of this problem, but they don't provide full centralization of all applications and related data objects (a brief overview of the advantages and disadvantages offered by these solutions will be included later in this chapter).

There is no platform that displays sanctioned SaaS applications and their related data objects as a collection of IT assets (like management configuration items—CIs) as well as the descriptive and contextual relationships between such IT assets. To put it more simply, there is no way to get a holistic view of users, groups, files, third-party applications, and data in a SaaS environment. IT departments are effectively operating with blinders on.

Challenge #2: SaaS Sprawl Leads to a Loss of Administrative Control

The SaaS explosion means that IT professionals lose critical root access, which, in turn, leads to a loss of administrative control. Today, each SaaS vendor determines the level of access IT departments have to the application's data. Consequently, IT and security teams no longer natively have root access to users, data, files, settings, etc. That access is available purely via APIs or not at all. This inhibits their ability to even see what's in their SaaS environment, let alone troubleshoot or fully secure it.

IT departments are expected to control, operate, and manage their SaaS environments. Yet wrangling all this data is an onerous, if not impossible, task for multiple reasons:

- **The sheer amount of data is unmanageable, and it's always changing.** Depending on the size of the organization, IT departments must manage tens of thousands—perhaps millions—of files owned by employees. The onus is on IT to ensure every single file is shared correctly and no sensitive data is inadvertently exposed to the public. One inappropriately shared document or misconfigured setting could have disastrous consequences, like a data breach or a compliance violation (case in point: Boeing suffered a 36,000-employee data breach in early 2017 when an employee emailed

The onus is on IT to ensure every single file is shared correctly and no sensitive data is inadvertently exposed to the public. One inappropriately shared document or misconfigured setting could have **disastrous consequences**, like a data breach or a compliance violation.

a spreadsheet to his spouse asking for formatting help. The spreadsheet contained 36,000 workers' first and last names, dates and places of birth, employee ID data, and Social Security numbers).[10] The more data there is—the bigger that sprawl is—the harder it is to stay in compliance. The amount of operational data that SaaS applications generate is staggering as well. In G Suite, for example, an audit log entry is created every time a user views, creates, previews, prints, updates, deletes, downloads, or shares Drive content.[11] Similarly, Salesforce's event-monitoring API creates event log files for logins, logouts, API calls, Apex executions, report exports, Visualforce page loads, and more.[12] Multiply that by tens, hundreds, or thousands of users. Data begets data, creating an enormous information sprawl. Compounding this problem is that this data is "living." It is continuously being changed, deleted, and added to, which makes the data even harder to manage.

▲ **SaaS data is fragmented; it lives everywhere.** SaaS data is not centralized. It doesn't live on users' devices; it lives on a server somewhere, which IT has little visibility into. The issue is that data overlaps everywhere. For instance, Drive and Dropbox files can be shared in Slack. Depending on

10 McIntosh, Andrew. "Boeing discloses 36,000-employee data breach after email to spouse for help." Puget Sound Business Journal, https://www.bizjournals.com/seattle/news/2017/02/28/boeing-discloses-36-000-employee-data-breach.html. Accessed 18 July 2017.
11 "Drive audit log: View user Drive file activity." G Suite Administrator Help, https://support.google.com/a/answer/4579696. Accessed 15 June 2017.
12 "Event Monitoring." Salesforce, https://trailhead.salesforce.com/en/modules/event_monitoring/units/event_monitoring_intro. Accessed 17 July 2017.

the environment, a user can install third-party applications (like Chrome extensions, bots, or marketplace applications) to his heart's content. Many SaaS applications work together in some way, and all it takes is one seemingly innocuous application integration (like the 18F example) to constitute a data breach. What's more, data can live across multiple instances of the same SaaS application too.

- **There is no consistency across SaaS applications, nor is there a unified view.** In multi-SaaS environments, settings are adjusted in different places (e.g., an admin console vs. a Team Settings page). Data objects all have different nomenclatures (e.g., a "group" vs. a "channel"). Each application has its own ecosystem—its own data silo. There is no single unified view or administrative vocabulary across SaaS applications for IT professionals. Not only does this mean IT departments must visit multiple disparate admin consoles every day to carry out administrative tasks, but it also means they have no clarity over their environment. There is no clear-cut way to view everything. For example, IT cannot view detailed information on all of a user's files, file types, permissions, or memberships in one centralized place. This information is usually not organized that way. Training and keeping up with change is also more difficult.

Ultimately, this type of SaaS sprawl means IT has little to no control over what enters the environment. Without the ability to centralize and view its organization's data, IT lacks effective oversight. It impairs the department's capacity to understand what's even happening in the environment.

Without the ability to centralize and view its organization's data, **IT lacks effective oversight.**

The control IT has over its environment is tenuous at best. If organizations are not currently encountering these problems, then they eventually will. Given the rapid pace of SaaS sprawl, it is unavoidable.

The Centralization Landscape

As mentioned above, there are solutions that solve parts of the SaaS sprawl conundrum, but they do not centralize data fully in a SaaS environment. Here is a brief look at some of the advantages and disadvantages offered by these solutions:

VENDOR CATEGORY	CENTRALIZATION FUNCTIONALITY	ADVANTAGES	DISADVANTAGES
Cloud Access Security Brokers (CASBs)	Shows all shadow and IT-sanctioned applications being used by an organization Helps visualize user activity	Can “see” everything due to position in the network as a proxy	Proxy limits contextual information, classification, and representation into user/data grid views Requires VPN Disrupts end user experience at times Is a single point of failure for access to SaaS apps, given its inline implementation
Identity-as-a-Service (IDaaS)	Shows all users and the access they have been granted to applications that the IDaaS vendor supports Shows permissions each user has been granted and their memberships in some groups	Typically support a very wide range of applications Can show user and location activity “during login” for app access	Cannot centralize any data objects that users create inside applications No awareness of data objects or user activity “post-login” inside apps
SaaS Application Management and Security Platform	Shows all data objects and associated metadata living inside IT-sanctioned SaaS apps	Offers deep visibility into wide range of data objects (users, files, groups, permissions, etc.) and all associated metadata and activity “post login,” when the application is actually being used	Typically supports a narrow range of applications due to depth and complexity of integrations

The Solution: Centralization

To solve the challenges created by SaaS sprawl, the very first element IT departments needs is data centralization. This is the foundation of the framework.

Centralization refers to the state of having a complete, consolidated view of IT's systems of record. It means IT can view all of its data objects across one or multiple SaaS applications in one place. The entire framework relies on complete centralization of SaaS data and the web of data objects connected across applications. IT requires this clarity before it can do anything else to manage a SaaS environment.

But the challenge with SaaS data, as mentioned earlier, is that it's all dissimilar; files stored in file storage services differ from those in chat applications, which differ from those in CRM applications. Data is stored in different places, and each application has its own set of APIs for accessing that data programmatically. Each application has different access, permissions, and sharing models as well. Thus, for IT to have true Centralization, data objects like audit logs, calendars, files, metadata, groups, users (including external users), and tickets must be ingested and normalized across different SaaS providers.

Use Cases: How Centralization Enables SaaS Management Success

As the foundation of the entire framework, visibility into SaaS environments is critical. IT cannot control an environment it cannot observe, nor can it take any meaningful action unless it sees all the data centralized in one place first.

"If [IT departments] can't see these cloud services being consumed, they can't see the risk that's being incurred," writes CIO.com. "If you can't see it, you really can't manage it."[13]

Below are questions that IT should be able to answer about their environment (but in order to do so, SaaS data must be centralized):

---

13 Corbin, Kenneth. "CIOs vastly underestimate extent of shadow IT." CIO, http://www.cio.com/article/2968281/cio-role/cios-vastly-underestimate-extent-of-shadow-it.html. Accessed 15 June 2017.

- Which third-party applications have access to the organization's data? What kind of access do they have?
- What kind of admin privileges do employees have across SaaS applications?
- What is each user and/or administrator doing in each SaaS application?
- What kind of member access, sharing options, and settings do the organization's groups have (e.g., who can view the group, members, and member email addresses; who can add/remove people; who can post topics, etc.)?
- How many calendars are publicly shared?
- Who has super admin access, either for a single SaaS application or multiple applications?
- What are all the pages that exist on the domain (e.g., a company intranet, client project sites)?
- What automations/workflows are set up in a single SaaS application or across multiple applications?
- How can IT view and manage every user's files, users, and/or groups across all SaaS applications in one place?
- How can IT view contextualized event logs across SaaS applications?

Use Cases: From Routine to Enlightened

There are four "stages" that the use cases fall into: Routine, Status Quo, Transformative, and Enlightened.

Routine: Simple, mundane tasks that IT teams must do manually and/or one by one. They usually require one or two steps in a single native admin console.

Status Quo: Common tasks that most (if not all) IT teams do today. They are slightly more involved and may require multiple steps in a native admin console.

Transformative: Increasingly complex, time-consuming tasks that IT teams need to do if they are transforming in some way—e.g., scaling quickly, opening new global offices, being acquired or divested, working with large amounts of contractors, etc. Transformative organizations need to think differently about operational security, productivity, and efficiency as their operations evolve. These tasks may require multiple steps in multiple native admin consoles and/or additional third-party assistance.

Enlightened: Extremely complex, advanced tasks that often cannot be completed without the help of custom development or third-party platforms that enable cross-application workflows. These capabilities are highly sought after and represent the peak for IT teams in terms of operational security, productivity, and efficiency.

These use cases are plotted such that readers can get an idea of the complexity and effort needed to carry them out, as well as how much value they bring to IT departments.

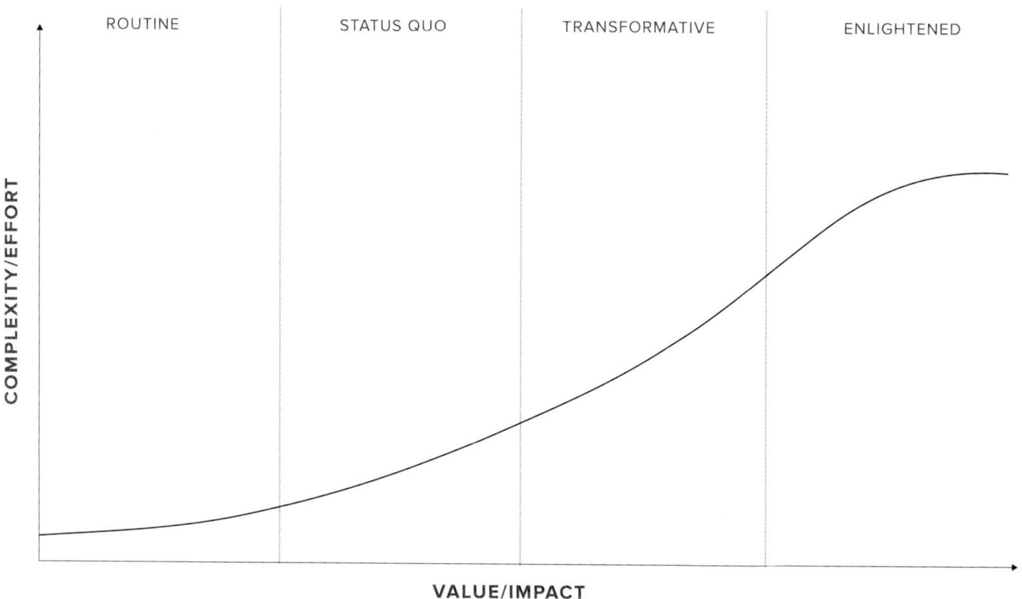

Figure 4

Here are several use cases for Centralization, plotted on the graph:

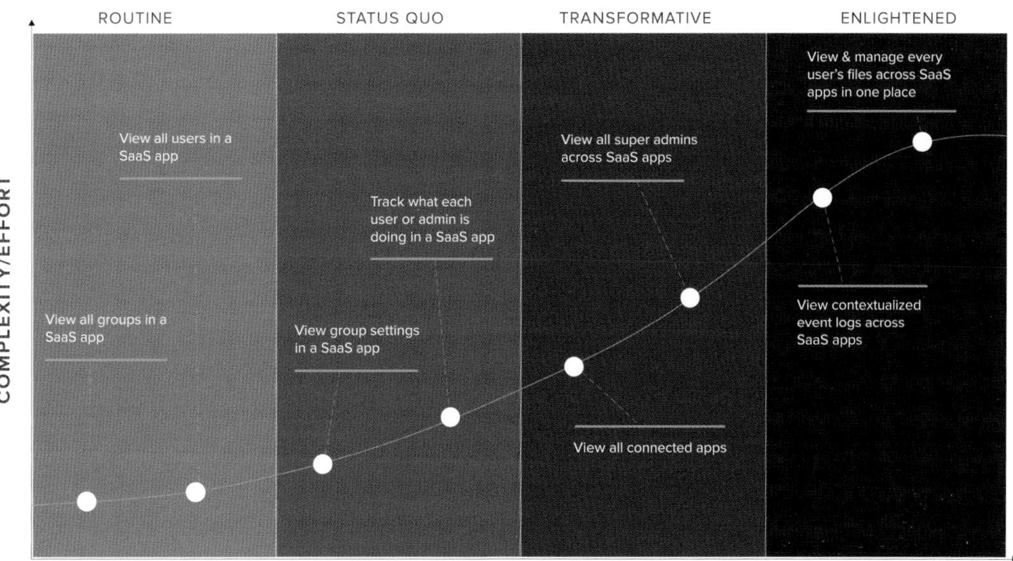

Figure 5

IT administrators can get visibility (albeit limited) in native admin consoles. They can see all their users and groups in one place. However, getting visibility into additional data objects is more challenging. Viewing all files, calendars, third-party applications, and external users, not to mention actually tracking what users are doing with these data objects, requires Centralization. Finally, viewing contextualized event logs across SaaS applications, as well as viewing and managing every user's files across SaaS applications in one place, are examples of highly complex tasks in the "Enlightened" stage. It is very difficult for IT to get this kind of centralized view, but it would be highly valuable. ▲

Next: Discoverability

Having centralized data lays the groundwork for the next steps in the framework. Once the sprawl of data objects has been consolidated and normalized, what's required next is a way to discover specific subsets of data.

CHAPTER TWO
DISCOVERABILITY

CHAPTER HIGHLIGHTS

DISCOVERABILITY

Even if SaaS data is centralized, a new problem soon becomes readily apparent: **IT cannot *find* any critical data amidst that sprawl**.

To secure its environment, IT needs to be able to locate files that have specific sharing settings, find documents and calendars that are publicly shared, identify external users who still have access to data, and identify third-party applications that have dangerously excessive access to data.

But native admin consoles **make it difficult, if not impossible, for IT to locate these data objects**.

The stakes are extraordinarily high. **Third-party applications can potentially access corporate data and privileged users**; competitors can view public data files and conduct industrial espionage; external users can abuse their access rights and steal data. The fallout from such incidents can be severe.

Thus, **IT must have the ability to "discover,"** filter, and find relevant data objects so that it can secure its SaaS environment.

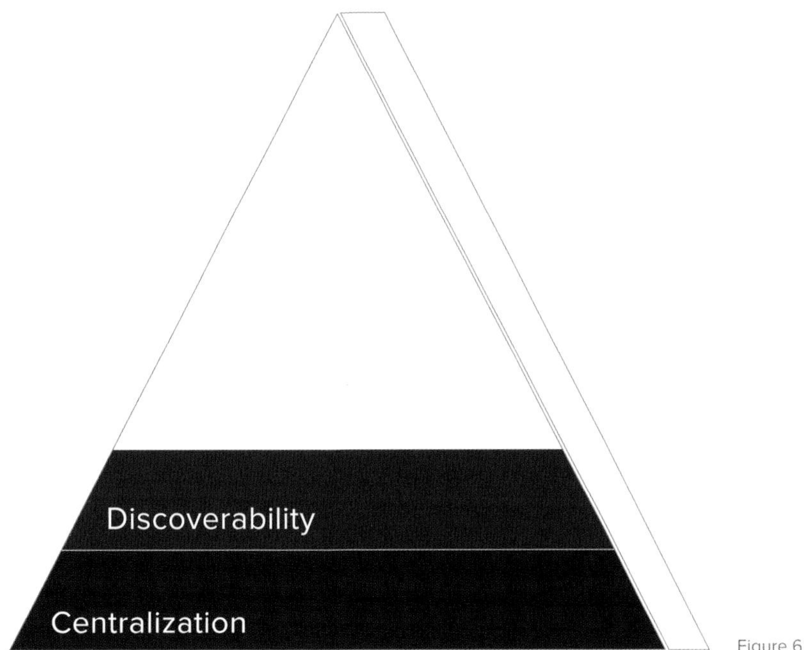

Figure 6

SaaS Application Management and Security Framework

DISCOVERABILITY

The Challenge: Finding Critical Information

Centralization is the first step to corralling SaaS sprawl. But even if SaaS data is consolidated in one place, a new problem soon becomes readily apparent: IT teams cannot find any critical data amidst that sprawl.

For example, an IT team may need to identify all users who meet certain criteria, or locate data contained in objects like files and calendars that may be shared publicly (intentionally or by accident). However, there is often no easy way to filter this information, especially across multiple attributes. Native admin consoles make it difficult, if not impossible, for IT to locate relevant data objects. It becomes a painstaking manual process, one that's often done in a spreadsheet. IT must examine each file, folder, mailbox, calendar, or user account, one by one, to see and resolve issues related to sharing settings, permissions, memberships, administrative privileges, third-party applications, and email rules (that is, if they can even see that data at all).

Files

This type of tedious work around files in particular results in substantial productivity loss. In a survey of more than 421 IT system administrators in the US, 17% of respondents said they spent at least 20 hours weekly on resolutions for unapproved file sharing and other file storage obstacles, resulting in a $34,450 productivity loss each year.[14] The same survey found that businesses can waste as much as $154,954 each year in productivity and ticket costs related to unauthorized cloud services, lost or inaccessible files, and other file sharing issues.[15]

Additionally, without the ability to filter and find data easily, IT has no way of knowing if its organization is in compliance with industry regulations or internal policies. According to a Ponemon Institute study, "Despite the risk, 64% of respondents say their organizations are in the dark about whether or not file-sharing activities are in compliance with laws and regulations. They have not conducted audits or assessment in the past 24 months. Only 9% of respondents say their organizations are certified and fully compliant today with ISO 27001 (the international standard for process-based security)."[16]

Almost one-third of respondents say more than half of the employees in their organizations regularly share files outside the company/beyond the firewall. 16% cannot determine this.

- Ponemon Institute

14 "Survey: Nearly $160,000 in Productivity Hours Lost Annually on File Storage, Sharing, Access." Business Wire, http://www.businesswire.com/news/home/20161004005410/en/Survey-160000-Productivity-Hours-Lost-Annually-File. Accessed 20 June 2017.

15 Ibid.

16 "Breaking Bad: The Risk of Unsecure File Sharing." Ponemon Institute, https://www.intralinks.com/platform-solutions/solutions/via/breaking-bad-risk-unsecure-file-sharing. Accessed 7 July 2017.

Similarly, almost one-third of respondents say more than half of the employees in their organizations regularly share files outside the company/beyond the firewall. 16% cannot determine this.[17] While it might be disconcerting that so many employees regularly share files outside the company, what's even more alarming is that some IT practitioners cannot even ascertain this percentage at their organizations. This 16% of respondents lack Discoverability. They cannot find crucial data amidst the sprawl, and thus cannot secure their environment effectively.

Calendars

Another area where Discoverability is critical is calendars. Calendars can contain a shocking amount of sensitive information between event details, attached files, and more. And for instance, if an employee's Google calendar is shared publicly, it will be indexed by Google, meaning it will appear in public search results. This can be disastrous. All of a user's calendar information, including event details, will be available to anyone. For example, one need only type "site:calendar.google.com" + "any search term" into a browser to find public Google calendars. This means that anyone on the Internet can potentially view sensitive details on corporate calendars that are inadvertently made public.

Consulting firms made headlines a few years ago when they ran into this exact situation. At McKinsey, corporate data slipped out via Google Calendar when the dial-in number and passcode for a weekly internal communication meeting was found just by searching Google. Details for several JPMorgan Chase & Co. conference calls relating to the company's storage systems could also be seen publicly, as well as dial-in information for compliance meetings at Deloitte.[18]

"This is pretty much exactly the kind of recon necessary to start doing industrial espionage," writes PCWorld. "Weekly meetings that discuss key internal information?

17 Ibid.
18 McMillan, Robert. "Corporate Data Slips Out Via Google Calendar." PCWorld, http://www.pcworld.com/article/130868/article.html. Accessed 6 July 2017.

Not looking good. Sometimes you see major leaks in the least likely places."[19]

Stories about industrial espionage abound. Here is one example: When a sales executive at a large pharmaceutical company recently noticed that all his clients kept canceling their meetings with him at the last minute, he pressed them for a reason. They admitted they had all set up meetings with the same competitor. He soon realized that his calendar, which included the set dates and times of client meetings, was shared publicly. The competitor had found his calendar online and was swooping in and stealing away clients, wooing them with a better deal.

Additionally, all it takes is a little clever social engineering and public calendar information to orchestrate a successful ruse that provides access to confidential systems or networks. "A search for 'routine maintenance' [on Google Calendars] produced some eyebrow-raisers. If you wanted to break into a company, what better way than to impersonate the repair guy? Worse yet, if a crook knows exactly when the repair guy is supposed to show up, he can call ahead and move up the appointment," writes the Washington Post.[20]

All it takes is a little clever social engineering and public calendar information to orchestrate a successful ruse that provides **access to confidential systems or networks**.

The perils associated with public calendars are endless. As a result, it is absolutely critical for IT administrators to have the ability to filter and discover these calendars in their organizations so that they can protect sensitive data and secure their environments.

---

19 Ibid.
20 Krebs, Brian. "A Word of Caution About Google Calendar." Washington Post, http://voices.washingtonpost.com/securityfix/2007/07/google_calendar_goofs.html. Accessed 7 July 2017.

External Users

One of the most dangerous security threats facing IT-reliant organizations today is from rogue contractors who have full or partial access to company data. SaaS applications make it particularly easy to create accounts for new users. The problem, however, lies in granting the right amount of privileges to external users who only need temporary, limited access (e.g., consultants, freelancers, contractors, seasonal workers who are employed for a short time).

Although most SaaS applications can limit the access users have in some way, these controls are rarely sufficient; they are not exact enough. Thus, IT has no choice but to grant contractors full access. There's often no way, however, to automate removal of that access, and large organizations have too many of these accounts to track manually. Contractors end up retaining full or partial access long after their contracts end, creating a major security risk. Therefore, one area where Discoverability is especially important is understanding and controlling external access.

Given that the freelancer economy is on the rise—40% of America's workforce will be freelancers by 2020[21]—managing external access correctly is an increasingly important challenge for IT. This is especially true given that freelancers may not be inclined to encrypt their laptops, password-protect their phones, etc.

It's "imperative for organizations to pay attention to contractors, but the reality is it's no easy task," writes SC Media, a cybersecurity source. "What makes this area of data security such a challenge is finding the right balance between limiting risk and opening up access to sensitive applications and data that a contractor needs to perform their job."[22]

External users are also especially prone to slipping through the cracks, which makes it all the more important that IT departments be able to filter and see data about them. "Oftentimes

21 Neuner, Jeremy. "40% of America's workforce will be freelancers by 2020." Quartz, https://qz.com/65279/40-of-americas-workforce-will-be-freelancers-by-2020/. Accessed 15 June 2017.
22 Trulove, Paul. "Data breach alert: the rising threat of contractors." SC Media, https://www.scmagazineuk.com/data-breach-alert-the-rising-threat-of-contractors/article/534744/. Accessed 7 July 2017.

[contractors] have access to very sensitive corporate data, including trade secrets, strategic plans, and other intellectual property. And an organization—in a rush to hire a third party for help on a project or simply shorthanded—might not have time for due diligence or the resources to provide adequate supervision. All that adds up to increased risk for either accidental loss or malicious theft of data," writes TechTarget.[23]

TechTarget continues: "Over time, continued reliance on an individual contractor will increase the risk to an organization and the consultant becomes more difficult to replace or terminate—a phenomenon that can be called 'dependency risk.' As the contractor becomes more entrenched, there is a tendency to provide less oversight."[24]

SaaS-Powered Workplaces (defined as organizations where 80% or more of their business applications are SaaS) experience these problems around external users very keenly. Research shows that they are 2.9 times more likely than average workplaces to say that understanding external access is a challenge.[25] These insights from IT professionals running SaaS-Powered Workplaces offer a prescient glimpse into the future. What this indicates is that most SaaS administrators cannot easily find information around external access (e.g., which external users no longer require access to data, yet still have it). Additionally, it indicates that administrators cannot adequately supervise or limit external users' activity (e.g., grant limited privileges, auto-expire their access rights after a few weeks, closely monitor their activity, etc.).

This problem only worsens as companies invariably scale and adopt more SaaS applications. In a survey of more than 700 Fortune 1000 IT professionals, 48% of survey respondents said third-party contractors represented the biggest threat in their environment.[26]

Furthermore, research from PwC reveals that contractors account for 18% of the most serious

---

23 "Are you putting information at risk by using contractors?" TechTarget, http://searchsecurity.techtarget.com/magazineContent/Are-you-putting-information-at-risk-by-using-contractors. Accessed 7 July 2017.

24 Ibid.

25 "2017 State of the SaaS-Powered Workplace Report." BetterCloud, https://www.bettercloud.com/monitor/state-of-the-saas-powered-workplace-report. Accessed 15 June 2017.

26 "Insider Threats Get More Difficult To Detect." Dark Reading, http://www.darkreading.com/risk-management/insider-threats-get-more-difficult-to-detect/d/d-id/1111712?print=yes. Accessed 7 July 2017.

> Cumulatively, security incidents stemming from negligent and careless employees or contractors cost the most money: organizations spent about **$2.3 million annually** dealing with the fallout from such incidents, at an average of about **$207,000 per incident**.
>
> - Ponemon Institute

breaches in UK firms of varying sizes.[27] If external access is not controlled, the financial ramifications can be exorbitant. Cumulatively, security incidents stemming from negligent and careless employees or contractors cost the most money. According to a Ponemon Institute survey, organizations spent about $2.3 million annually dealing with the fallout from such incidents, at an average of about $207,000 per incident.[28]

External users can steal data or compromise systems for any number of reasons: greed, malice, personal disagreements, financial motivations, revenge, whistle-blowing, etc. It is easy for them to go rogue if they retain too much access for too long. Consider the case of Edward Snowden, a contractor who used his enhanced privileges as a system administrator to exploit a security hole and take millions of classified documents from the NSA's servers—earning him both "hero" and "traitor" labels.

Third-Party Applications

There is no doubt that third-party applications (like Chrome extensions or Slack apps) can be useful. They help people work more efficiently and boost productivity. However, their outward simplicity belies their dangers. Because they are very simple to install—it usually

27 Trulove, Paul. "Data breach alert: the rising threat of contractors." SC Media, https://www.scmagazineuk.com/data-breach-alert-the-rising-threat-of-contractors/article/534744/. Accessed 7 July 2017.

28 Vijayan, Jay. "Insider Incidents Cost Companies $4.3 Million Per Year On Average." Dark Reading, http://www.darkreading.com/vulnerabilities---threats/insider-threats/insider-incidents-cost-companies-$43-million-per-year-on-average-/d/d-id/1326891. Accessed 15 June 2017.

> "These types of apps can request permission to **view, delete, transfer,** and **store corporate data** when enabled using corporate credentials, which is why it's important that organizations identify these applications, particularly those that pose the highest risk."
>
> - Security Week

just takes one or two clicks—users can often install them on their own without IT's knowledge or approval. In doing so, users can unknowingly grant these applications access to sensitive data.

OAuth-connected applications in particular pose a security threat because they can have extensive access to corporate data. "These types of apps can request permission to view, delete, transfer, and store corporate data when enabled using corporate credentials, which is why it's important that organizations identify these applications, particularly those that pose the highest risk," writes Security Week.[29]

When employees install applications and authorize OAuth scopes via their corporate credentials, it can lead to serious problems. "If these apps are malicious by design, or the connected application's vendor is compromised, this opens the door to cybercriminals deleting accounts, externalizing or transferring information, provisioning or deprovisioning users, changing users' passwords, modifying administrator's settings, performing email log searches, and more," according to Security Week.[30]

Example: Chrome Extensions

In particular, Chrome extensions can be a major threat to corporate security.

29 Kovacs, Eduard. "Enterprises Warned About Risky Connected Third-Party Apps." Security Week, http://www.securityweek.com/enterprises-warned-about-risky-connected-third-party-apps. Accessed 12 July 2017.
30 Ibid.

"Because [Chrome] extensions add extra functionality to the browser, they need a lot of power. Extensions often ask for a variety of permissions that come from Chrome's APIs. For example, extensions can intercept web requests from the browser and modify traffic and inject JavaScript into web pages. That is just way too powerful. We need to be way more careful into which extensions we give these permissions," writes InfoWorld.[31]

An analysis by security researchers of 48,000 Chrome extensions uncovered 130 outright malicious extensions and 4,712 suspicious ones, which engaged in a variety of affiliate fraud, credential theft, advertising fraud, and social network abuse.[32] The actions these extensions took were mostly undetectable to regular users.

There is little standing between a user and a nefarious Chrome extension. All it takes is one user breezily hitting "Accept" on a malicious Chrome extension to wreak serious havoc in an organization. In fact, "with the right permissions, your new Chrome add-on could steal your user credentials, post as you on social media, read your emails, help launch a DDoS attack, and more," writes BetaNews.[33]

"[Chrome] extensions can intercept web requests from the browser and modify traffic and inject JavaScript into web pages. That is just way too powerful. **We need to be way more careful into which extensions we give these permissions.**"

- InfoWorld

In 2014, researchers at Carnegie Mellon University released an academic paper on the dangers posed by Chrome extensions, writing: "[Chrome] extensions can potentially make network requests, access the

31 Kirk, Jeremy. "Many Chrome browser extensions do sneaky things." InfoWorld, http://www.infoworld.com/article/2608717/web-browsers/many-chrome-browser-extensions-do-sneaky-things.html. Accessed 10 July 2017.
32 Ibid.
33 Williams, Mike. "How you might get hacked by a Chrome extension." BetaNews, https://betanews.com/2016/07/27/how-you-might-get-hacked-by-a-chrome-extension/. Accessed 15 June 2017.

local file system, get low-level information about running processes, etc. [...] We demonstrate a series of attacks by which extensions can steal data, track user behavior, and collude to elevate their privileges. Although some attacks have previously been reported, we show that subtler versions can easily be devised that are less likely to be prevented by proposed defenses and can evade notice by the user."[34]

From a security standpoint, it is imperative that IT has the ability to filter and find third-party applications in a SaaS environment. The stakes—access to corporate data and privileged users—are extraordinarily high.

Example: Slack Apps and Dropbox Apps (Varying Scopes)

As with Chrome extensions, users can easily install apps that integrate with Slack. These can help with general productivity, department-specific needs (e.g., customer service apps), or with automated tasks (e.g., Slack bots). But Slack apps may have more access than people realize. They request various scopes, which grant different levels of access. For example, these are the classes of action in Slack, which become increasingly far-reaching:

- ▲ **Read:** Reading the full information about a single resource.
- ▲ **Write:** Modifying the resource in any way e.g. creating, editing, or deleting.
- ▲ **History:** Accessing the message archive of channels, DMs, or private channels.
- ▲ **Identify:** Allows applications to confirm your identity.
- ▲ **Client:** Allows applications to connect to Slack as a client, and post messages on behalf of the user.
- ▲ **Admin:** Allows applications to perform administrative actions (requires the authorized user to be an admin).[35]

---

34 Bauer, Cai, Jia, Passaro & Tian. "Analyzing the Dangers Posed by Chrome Extensions." Carnegie Mellon University, Singapore Management University, and Institute for Infocomm Research. https://www.ece.cmu.edu/~lbauer/papers/2014/cns2014-browserattacks.pdf. Accessed 10 July 2017.

35 "OAuth Scopes." Slack, https://api.slack.com/docs/oauth-scopes. Accessed 10 July 2017.

As a result, while a Slack app may only be needed for certain limited use cases, it could ostensibly access entire archives of channels, DMs, or private channels; post messages on behalf of a user; or even perform administrative actions.

...while a Slack app **may only be needed for certain limited use cases**, it could ostensibly access entire archives of channels, DMs, or private channels; post messages on behalf of a user; or even perform administrative actions.

Similarly, when a user links an application to Dropbox, the application will request a specific level of access. This can include "access to specific types of folders and files anywhere in your Dropbox" (the application can access general groups of files, such as photos or documents, or specific kinds of files) to "access to *all* folders and files in your Dropbox" to "access to the email address associated with your account."[36]

Example: Other Cloud Applications

Third-party applications can also mean mobile applications, like games. This can be particularly problematic if users recycle their work login credentials for these applications.

"Cloud-based apps often gain access to the camera, location, data, and contacts on your phone. So you never know how much sensitive company information they may be snaffling. We could be giving hackers, fraudsters, and spies the keys to our company's back door, particularly if we naively use the same login details for external apps as we do for internal work apps," writes the BBC.

Additionally, with the rise of Bring Your Own Device (BYOD) policies in workplaces, there is

36 "What information can a third-party app access when I link it to my account?" Dropbox, https://www.dropbox.com/help/security/third-party-apps. Accessed 10 July 2017.

“Cloud-based apps often gain access to the camera, location, data, and contacts on your phone. So you **never know how much sensitive company information they may be snaffling.**"

- BBC

a heightened security risk because mobile applications can come loaded with malware. According to the BBC, these are monetized by selling users' information and phishing for banking credentials.[37] The worry for IT departments is that these third-party applications may not have strong security protocols in place because many are developed primarily for consumers.

“It's a mission-critical problem if you don't know which third-party apps have access to your data,” writes the BBC.[38] Indeed, IT can have all its SaaS data centralized but if they lack the tools to find critical data, they are virtually powerless. If they don't know which applications or bots are connected to their organization, which files and calendars are shared publicly, or which external users have access to sensitive data, then this knowledge gap puts their organization at great risk. Centralization does little good unless IT has the ability to filter and find important data objects.

The Discoverability Landscape

There are some vendors that do provide root access and enable IT to find data objects in their environment to some degree. However, they do not provide IT much depth beyond that. They do not afford IT the ability to extensively drill down and locate data objects based on specific attributes. Here is an overview of some of the advantages and disadvantages of various types of vendors in this category:

37 Wall, Matthew. “Is that app you're using for work a security threat?” BBC, http://www.bbc.com/news/business-37541594. Accessed 10 July 2017.
38 Ibid.

VENDOR CATEGORY	DISCOVERABILITY FUNCTIONALITY	ADVANTAGES	DISADVANTAGES
CASBs	API-mode CASBs provide rudimentary search functions for a small set of data objects	Good for basic rudimentary user activity discovery	Cannot provide comprehensive discoverability across any data type or app
IDaaS	Provides user and group discovery for IT-sanctioned applications	Good for user and group information discovery	Only syncs in directory data (specifically identity-related data like users and groups)
SaaS Application Management and Security Platform	Extensive variety of canonical data object models, along with normalized, indexed, and structured data across disparate applications makes filtering, searching, and discovery fast and flexible	Native app awareness directly via API connectivity provides the best way to view and search aggregate content across applications	Every target application has to be integrated into the platform with full coverage for all data object models within the application

The Solution: Discoverability

Discoverability takes Centralization one step further: It makes sense of the sea of detailed information that has been consolidated into one place. It's the ability to search, sort, and filter data in order to find specific data objects that are important to IT. Essentially, Discoverability is about taking a massive data set and pinpointing the data that is purely relevant to a specific management or operational task. It's executed at an organization level or at a user level. Typically, the most common use case for Discoverability is to be able to see all data objects, settings, and activity tied to a specific user across all their SaaS applications.

There can be dozens, hundreds, even thousands of users in an environment. Those users may have thousands, even millions, of files across SaaS applications. Similarly, there may be hundreds of channels and groups across SaaS applications. Some of them may contain external members. Some may contain internal members who should no longer belong because they've moved departments. It's impossible to break down that data unless IT can perform complex queries against the database(s) to return specific results in real time. To discover relevant data points, IT needs the ability to interact with the database via filters of some sort, in order to find objects based on their attributes.

For example, if IT wanted to identify spreadsheets across all SaaS applications that have been shared externally, it would first need the ability to filter down to files, then filter down to spreadsheets specifically, then filter down to spreadsheets that have been shared externally.

Achieving Discoverability requires scanning multiple data tables in multiple databases. For instance, files may be stored in one table along with their associated metadata and permissions (e.g., who the file owner is, and what the file's visibility is). In order to produce relevant results, IT might have to query across multiple databases to find relationships between data objects and their associated metadata. For example, a query might search for all spreadsheets that are shared publicly *and* are owned by users on the HR team *and* contain the word "Confidential" in the content. Another query might search for all secondary calendars that are shared publicly *and* are owned by users on the sales team.

It's also important to note that various filters require the use of different types of databases, each of which must be optimized for each specific use case. For example, when IT is filtering for users, the database must be highly optimized for a high volume of searches. On the other hand, a database that houses files may not be quite as performant in terms of queries, but must be able to handle massive datasets. For maximum efficiency, each database must be optimized for the type of filtering that is taking place.

Use Cases: How Discoverability Enables SaaS Management Success

Unless IT teams have Discoverability in place, they cannot answer the following critical questions about their environment:

- Can all publicly shared documents be identified?

- ▲ Can all spreadsheets (either within one SaaS application or across multiple SaaS applications) that have been shared outside an organization and contain protected health information (PHI) or personally identifiable information (PII) be identified?
- ▲ Can anyone sharing files with their personal email accounts be identified?
- ▲ Can all external users (e.g., consultants, contractors) across SaaS applications be identified?
- ▲ Can external users who are no longer under contract (but still have access to specific applications and data) be identified?
- ▲ Can concurrent failed logins across SaaS applications be easily identified?
- ▲ Can the most active users by public file sharing behavior across all applications be identified?
- ▲ Can permission levels by departments be filtered? (e.g., perhaps the sales department should have the ability to share documents externally, but finance shouldn't)
- ▲ Can anyone who has (or doesn't have) two-factor authentication turned on be identified?
- ▲ Can IT determine if and when super admins are added to or removed from each SaaS application?
- ▲ Can IT ensure that specific (e.g., company-wide) groups have the correct settings (e.g., they are not public on the Internet, and have the correct admins in place)?
- ▲ Can IT ensure that any high-risk third-party applications are not connected?
- ▲ Can any groups that meet a certain exposure criteria (e.g., public on the Internet) be identified?
- ▲ Can all publicly shared calendars be identified?
- ▲ Can any employees who are forwarding any emails to personal email

accounts or other external domains be identified?

- ▲ Can employees who have connected more than two mobile devices (thus violating corporate policy, for example) be identified?
- ▲ Can employees belonging to the wrong groups be identified?
- ▲ Can employees who belong to private groups or channels be identified?
- ▲ Can departed employees who still own recurring events on their calendars be identified?
- ▲ Can the level of access third-party applications have to an organization's data (e.g., read/write permissions across multiple applications) be identified?
- ▲ Can employees who are delegated administrative access be identified?
- ▲ Can automations/workflows that touch users, files, etc. be identified?
- ▲ Can the channels or groups that have external users be identified?
- ▲ Can users whose cloud storage files are being mass-encrypted at high speeds, and thus likely infected with ransomware, be identified?

Discoverability is integral to operational security. Having the ability to locate and filter specific data is particularly critical from a security and compliance standpoint. Without the ability to discover at-risk users or files, there is little chance of remaining compliant or secure.

Use Cases: From Routine to Enlightened

Figure 7 shows several Discoverability use cases on the same complexity/effort vs. value/impact graph as in the previous chapter.

There are two examples of "Routine" Discoverability use cases here: Searching for a user in a SaaS application, and identifying who is not enrolled in two-step verification. This information is easily discoverable in a native admin console, yet brings little value to IT since it's quite broad.

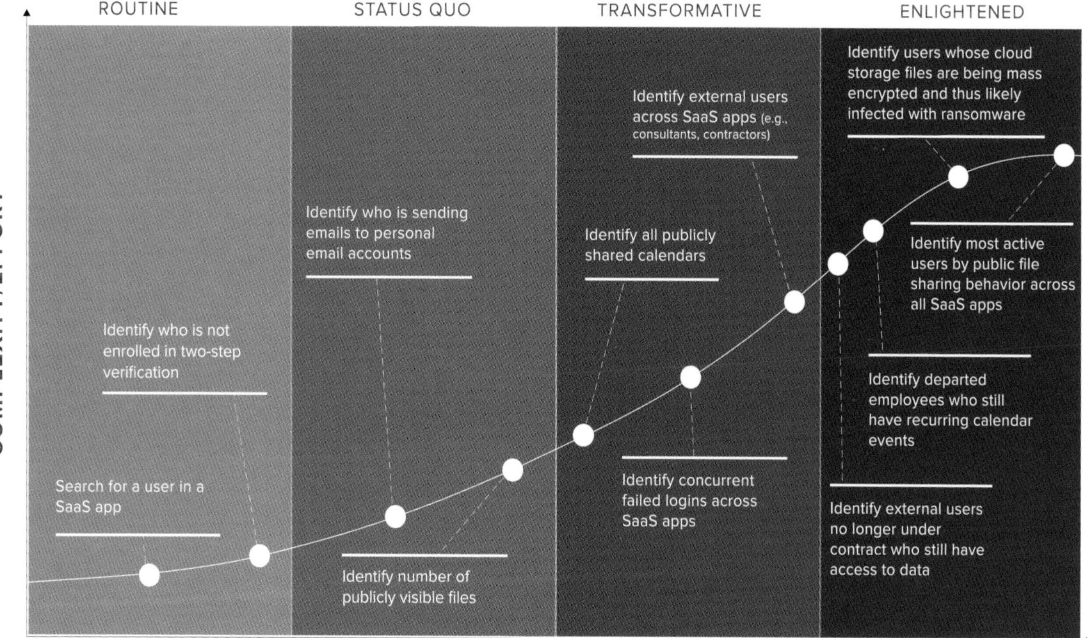

Figure 7

However, as one moves along the curve, these tasks become more complex and more valuable. Identifying concurrent failed logins across SaaS applications, for instance, is important for security purposes but is a time-consuming, multi-step process. IT must log into each native admin console separately and manually match up the times, locations, etc. of multiple failed logins to identify concurrent events.

Finally, in the "Enlightened" stage, there are use cases like identifying external users (e.g., consultants) who are no longer under contract yet still have access to company data. IT cannot easily filter or find this type of data, yet this information is critical in order to maintain the operational security of an organization. ▲

Next: Insights

Centralization and Discoverability focus on pulling information from SaaS applications. But the next element, Insights, is about having information pushed as changes are made.

CHAPTER THREE

INSIGHTS

CHAPTER HIGHLIGHTS
INSIGHTS

IT cannot fix what it doesn't know is wrong.

IT cannot extract critical information about its environment. Alerts are often false alarms. Trying to divine meaning from millions of audit logs is futile. **Finding an event that is actually relevant and critical is like finding a needle in a haystack.**

Take Target's massive 2013 data breach as an example. Target's security team had actually reviewed—and ignored—urgent warnings from a threat-detection tool about unknown malware on its network, because **the alerts were so common**.

With the proper insights, **IT can ensure compliance with corporate policies** as well as industry and government regulations, correlate high-risk activities, prevent unauthorized data access, measure user engagement, and recoup license costs.

What makes alerts valuable is the **ability to set custom triggers, thresholds, and windowing functionality,** as well as correlate disparate, high-risk activities.

Insights are an effective way to bolster operational security by remedying the **"You don't know what you don't know" challenge**. They help fill gaps in IT's knowledge about its SaaS environment, essentially providing the information IT didn't even know it was missing.

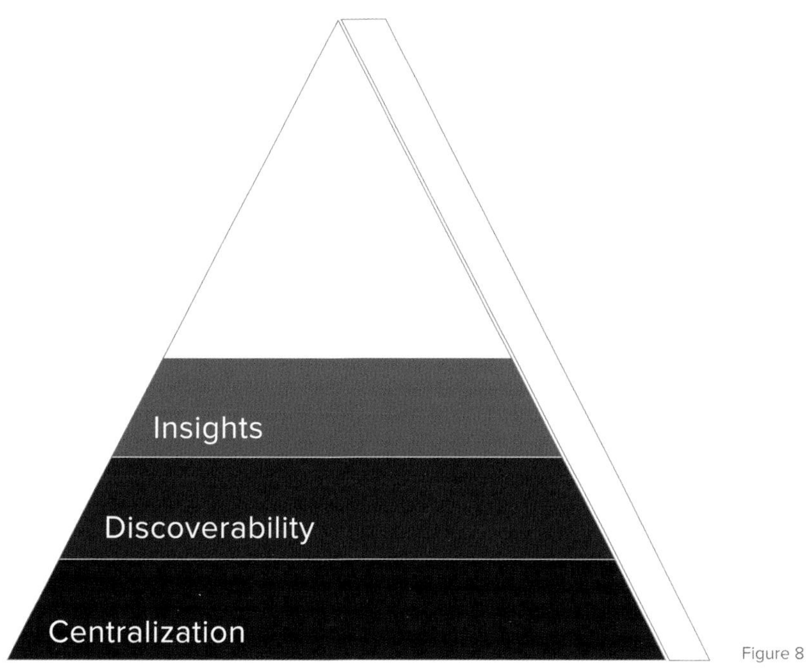

Figure 8

SaaS Application Management and Security Framework

INSIGHTS

Challenge #1: Combating Alert Fatigue and Finding the Needle in the Haystack

IT is responsible for responding quickly to incidents, data exposures, and security risks before they become critical issues. Therefore, IT needs the ability to extract vital insights about users, groups, and data in its SaaS applications in order to stay on top of emerging security threats. However, this is natively impossible. IT has limited insight into the important events going on within its SaaS applications.

“This [SaaS] data ‘sprawl factor’ compromises any organization’s ability to gather proper, trustworthy insight that can help it not only understand, but also manage corporate performance effectively," writes BetaNews.[39]

At the heart of SaaS sprawl is a disconcerting issue: IT cannot fix what it doesn't know is wrong. As the adage goes, "You don't know what you don't know."

It's not for lack of trying. IT departments receive dozens of notifications, usually by email. But noisy alerts are one of IT's biggest bugbears. Alerts are often false alarms—irrelevant, dismissible, excessive—and impossible to keep up with. Additionally, trying to divine meaning from millions of audit logs is a futile endeavor. Finding an event that is actually relevant and critical is like finding a needle in a haystack. According to CloudExpo Journal, "Large IT organizations can receive up to 150,000 alerts per day from their monitoring systems. How are IT employees supposed to sort through them all to pick out the one or two legitimate threats? It's simple—they can't."[40]

"Large IT organizations can receive up to **150,000 alerts per day** from their monitoring systems. How are IT employees supposed to sort through them all to pick out the one or two legitimate threats? It's simple—they can't."

- CloudExpo Journal

This endless stream of notifications creates alert fatigue, a mental state that is both overwhelming and dangerous. Administrators become desensitized and learn to ignore alerts. Security threats go unnoticed or undetected.

Take Target's massive 2013 data breach as an example. Target's security team had actually reviewed—and ignored—urgent warnings from a threat-detection tool about unknown

39 Peirce, Darren. "How enterprises can overcome SaaS' data fragmentation challenge." BetaNews, https://betanews.com/2017/02/17/saas-enterprise-data-fragmentation/. Accessed 15 June 2017.

40 McAlpin, Troy. "Biggest Fears of the Modern IT Manager." CloudExpo Journal, http://cloudcomputing.sys-con.com/node/3320783. Accessed 20 June 2017.

malware on its network[41] because the alerts were so common. Target's security team "incorrectly deemed the event log message a false positive. Instead of alerting management that the company was under attack, everyone remained silent as the logs filled up with evidence of the infiltration. This single bonehead move cost Target hundreds of millions of dollars, forced the resignation of the CEO and CIO, and eroded customer trust in the brand."[42] Target ended up settling with several US banks for $39 million, Visa for $67 million, and customers for $10 million.[43]

Additionally, a 2015 survey of 600+ IT professionals found that the number one pain point of being on-call was "alert fatigue—constantly being paged for non-actionable alerts."[44] Large organizations using G Suite can easily receive 1,000+ suspicious login alerts a day. The onslaught of alerts can paralyze IT, preventing them from taking meaningful, strategic actions. If everything is "important," nothing is.

Challenge #2: Receiving Context and Having Remediation Options

SaaS alerts are typically in a vacuum—that is, they lack context (i.e., structured metadata). This absence of context is problematic.

"SaaS metadata, more so than traditional metadata, contextualizes data for humans as well as machine processes [...] This means understanding metadata is vital for SaaS admins," writes Spanning. "For SaaS applications that are messaging and collaboration tools, metadata plays a vital role in enabling collaboration and control, as it contains information about sharing settings, labels, tags, and ownership."[45]

---

41 Schwartz, Mathew J. "Target Ignored Data Breach Alarms." DarkReading, http://www.darkreading.com/attacks-and-breaches/target-ignored-data-breach-alarms/d/d-id/1127712. Accessed 15 June 2017.
42 Grimes, Roger A. "10 security mistakes that will get you fired." InfoWorld, http://www.infoworld.com/article/2846758/security/10-it-security-mistakes-that-will-get-you-fired.html. Accessed 23 June 2017.
43 "Target settles for $39 million over data breach." CNN, http://money.cnn.com/2015/12/02/news/companies/target-data-breach-settlement/. Accessed 15 June 2017.
44 "2015 State of On-Call." VictorOps, https://victorops.com/wp-content/uploads/2015/12/VictorOps-InfoGraphic-Final-2.pdf. Accessed 15 June 2017.
45 "What is SaaS Metadata, and Why Do You Need to Protect It? Part I." Spanning, https://spanning.com/blog/what-is-saas-metadata-and-why-do-you-need-to-protect-it-part-i/. Accessed 15 June 2017.

"For SaaS applications that are messaging and collaboration tools, **metadata plays a vital role** in enabling collaboration and control, as it contains information about sharing settings, labels, tags, and ownership."

- Spanning

For example, IT might receive an alert stating that a user has been granted admin privileges, but this alert provides no context. IT doesn't know if this is a perfectly legitimate change or an urgent security risk. IT doesn't know which business unit or OU the user belongs to. Without any context, generalized alerts are meaningless. It makes it difficult, if not impossible, to discern what's actually critical. There is no context describing how or why violations occur.

Additionally, there are no remediation options for SaaS applications. An alert notifies IT teams when something is happening, but that information is not converted into anything actionable. IT does not know what the best remediation actions are. Without any action tied to it, there is little to no value to the alert. Indeed, in a 2016 survey of 800+ on-call professionals, 39% of respondents said lack of remediation information was a problem.[46]

Challenge #3: Receiving Cross-Application Insight

Native administration becomes unscalable as organizations adopt more SaaS products. Data resides in disparate applications across multiple consoles. For organizations that have adopted multiple SaaS applications, there is no way to centralize alerts across them or see everything at once; they are siloed. IT must check each application separately to see alerts, and some applications do not even provide alerts natively. Each

46 "2016 State of On-Call." VictorOps, https://victorops.com/wp-content/uploads/2016/12/VictorOps-State-of-On-Call-2016-2017-Report.pdf. Accessed 15 June 2017.

application also has its own alert types, delivery methods, terminology, and admin interface. There is no UI consistency across admin consoles. This disparity means that IT must train new IT hires on multiple admin consoles, which can be a massive onboarding time-suck.

Finally, cross-application insight does not exist. Alerts for one individual SaaS application might be useful, but what's immensely more valuable to IT is being able to bring cross-application correlations or anomalies to the surface. For example, intelligence around aberrant activity during multiple logins (or mass downloading events across multiple applications) is more meaningful to IT than intelligence around an isolated suspicious login. Why? Because these correlations can signify serious security problems or impending events, like employee resignation.

CSO reported a case where sales reps were quitting unexpectedly and taking data to their new jobs. Security analysts manually examined behavior patterns for employees, and then matched them with the behaviors of those who ultimately quit. They were able to correlate abnormal behaviors such as: completing mass exports of lead information, entering parts of the system where they didn't usually go, changing object information, deleting items, and doing any of these things from home or in the office on a Saturday afternoon.[47]

"With these early warning indicators, IT staff was able to put controls in place to stop massive downloads before they happened or freeze accounts for several hours until a manager had a chance to speak with the employee," writes CSO.[48]

If IT teams could receive insights into correlations between abnormal behavior and early warning signs across applications, they could easily minimize data theft, security risk, and business disruption. However, there is no way of obtaining these insights using native admin consoles. There is also no way to correlate events across applications because there is no common denominator across various vendors' activity logs.

---

47 Collett, Stacy. "Five signs an employee plans to leave with your company's data." CSO, http://www.csoonline.com/article/2975100/data-protection/five-signs-an-employee-plans-to-leave-with-your-companys-data.html. Accessed 19 July 2017.
48 Ibid.

The Insights Landscape

Again, while there are some vendors that do provide alerting capabilities, where they fall short is in providing contextual details; they do not provide context into assets that trigger alerts. Without context, IT does not have enough information to remediate incidents effectively.

Here is an overview of the advantages and disadvantages that some of the vendors in this category provide:

VENDOR CATEGORY	INSIGHTS FUNCTIONALITY	ADVANTAGES	DISADVANTAGES
CASBs	Insights are centered around data loss prevention and malware threat prevention	Good for pure security use cases to protect organizations from external threats and enable regulatory compliance	Not useful for day-to-day operational incidents that could occur from accidental user behavior Remediation (app block/user quarantine) disrupts user experience Is a single point of failure when deployed in proxy mode
Security Information and Event Management (SIEMs)	Able to normalize and store a variety of event logs Insights are obtained by building custom regular expression search queries across critical events	Flexibility to build custom and powerful search queries against the various event and data types ingested by SIEMs across disparate SaaS apps	Supports maximum two to three apps No prebuilt set of insights or alerts No remediation actions available for insights or incidents natively
SaaS Application Management and Security Platform	Has extensive methods of ingesting native SaaS application events and user activity and can normalize and store common canonical data objects across disparate SaaS apps Both pre-built and custom insight/alert definitions are available	Deep application awareness and contextual correlation of user and data activity provides for an operational security and service intelligence approach to surfacing critical app insights API-only architecture is best suited for native app knowledge to capture anomalous user behavior with an "outside-in" security model	Functionality is reliant on the number of application APIs that are exposed for events and user activity Real-time effectiveness and monitoring for insights are dependent on the nature of notification APIs (webhooks, callbacks, etc.) for applications

The Solution: Insights

Alerts need to be reliable and equipped with meaningful contextual details. Furthermore, they need to cut through the noise and reach the right IT leaders quickly. To combat the problems described earlier, the key is relevance: quality over quantity. IT needs a system akin to a sieve: something that restricts massive amounts of noisy data and only lets critical insights—that is, relevant, actionable information—through. When this happens, alert fatigue is all but eliminated.

The only way to surface those salient insights, however, is to have Centralization and Discoverability already in place. Otherwise, IT is forced to look through every single alert and log into each of the SaaS platforms separately. There is no other way to consistently detect patterns at scale. When critical data is centralized in one place, it provides visibility and saves IT the effort of hunting for data within individual applications or parsing through dense audit logs.

IT needs a system akin to a sieve: something that restricts massive amounts of noisy data and only lets critical insights—that is, relevant, actionable information—through. When this happens, alert fatigue is all but eliminated.

Once data is centralized and discoverable, alerts can be created with enough granularity to guarantee relevance. To do that, IT needs a complex event-processing engine to sift through millions of file events, user events, group events, etc. in a very short period of time and make meaning out of them. It must be able to catch an event, run an analysis on it, enrich it with additional data from other databases, and then match it in the same process stream. Because it can analyze different events in-stream, it can make valuable correlations across events (e.g., if there is a suspicious login for a user in two or

Insights help fill gaps in IT's knowledge about its SaaS environment, essentially providing the information **IT didn't even know it was missing**.

more applications, then it can combine these events and trigger an alert).

This event-processing engine allows IT to configure complex alerts and be told when a specific combination of events is happening, either within a single SaaS application or across multiple SaaS applications. Disparate, high-risk activities can be correlated granularly. This is where the normalization from Centralization comes into play. Because there is a canonical user model that normalizes data across applications, it's easy to start surfacing insights downstream. Thus, IT can receive alerts (e.g., a user with no manager) that are SaaS provider-agnostic, or for only a specific SaaS provider, if they so choose.

For alerts to provide critical insight to IT, they should be generated based not only on what's happening, but also *how* it's happening—e.g., if there's a high event frequency within a specific time window. For example, a user downloads one file in a SaaS application—that in and of itself is uninteresting and insignificant. But if that user downloads 99 more files from multiple SaaS applications in the following hour, then that event is more suspicious—and, in turn, more important. Multiple SaaS events are correlated, producing one intelligent alert.

What makes alerts valuable is the ability to set custom triggers, thresholds, and windowing functionality. With these insights, IT can ensure that it will only be alerted for events that truly do warrant action or further investigation. The right information is pushed at the right time to the right people, through the right communication channel (e.g., a chat application, email, SMS).

Insights are an effective way to bolster operational security by remedying the "You don't know what you don't know" challenge. They help fill gaps in IT's knowledge about its SaaS environment, essentially providing the information IT didn't even know it was missing.

Use Cases: How Insights Enable SaaS Management Success

In the face of overwhelming SaaS sprawl, IT must have the ability to create order from disorder. It must be able to derive relevant, critical insights from millions of data points. The sooner IT teams are made aware of these insights, the faster they can remediate any issues. Below are some examples of use cases where these types of insights can harden an organization's security posture:

Ensure compliance with corporate policies, as well as industry and government regulations.

The list of industry and federal compliance regulations seems to be ever-growing. Examples include HIPAA (Health Insurance Portability and Accountability Act of 1996), SOX (Sarbanes-Oxley Act of 2002), FERPA (The Family Educational Rights and Privacy Act), and PCI DSS (Payment Card Industry Data Security Standard). To remain in compliance, organizations must protect sensitive information and follow required compliance procedures. Non-compliance can result in hefty fines, so the ability to receive relevant alerts can greatly reduce potential regulatory compliance risks and security threats. IT teams should be able to ensure compliance at the bare minimum, but ideally, Insights should empower them to be "better than best."

- ▲ IT receives an alert when a user forwards emails containing PHI to a personal email account.
- ▲ IT receives an alert when any user in the finance department shares files publicly.
- ▲ IT receives an alert when users connect more than two mobile devices to their SaaS applications and thus violate corporate policy.

Correlate high-risk activities.

Within any organization, there are users and departments who are more high risk than others. Examples of these include executives, super admins, finance, HR, or external users. Similarly, there are certain data objects that are high risk, such as confidential files or OAuth-connected applications installed organization-wide. Additionally, there are certain activities that are high risk, such as new super admins being added to applications, erratic login activity, or data being connected to external domains (e.g., Gmail/Yahoo accounts). The ability to receive an alert for a confluence of these events is highly valuable, particularly because they can be predictors of impending events or security issues.

- ▲ IT receives an alert that predicts an employee's resignation, because several warning signs are correlated (e.g., a user mass downloads records from Salesforce in the span of seven days, forwards emails to a personal email account, and logs in repeatedly from home on the weekends).
- ▲ IT receives an alert that a user in the finance department has set up an email forwarding policy and turned off two-factor authentication.
- ▲ IT receives an alert when executives are forwarding emails with large attachments to an external domain (Gmail, Yahoo, Outlook).
- ▲ IT receives an alert when a secondary calendar owned by HR is created and has a “Public” sharing setting.

Prevent unauthorized data access.

- ▲ IT receives an alert when login anomalies pass a certain threshold and deviate from the norm by a certain percentage (e.g., a user logs in outside of normal working hours from a new location on a device he doesn’t normally use and uses applications he doesn’t typically need for his job function).
- ▲ IT receives an alert when a user installs a third-party application that hasn’t been reviewed yet and violates security criteria.
- ▲ IT receives an alert when a new super admin is added to an application.
- ▲ IT receives an alert when a user modifies group settings and changes them to public (or “Anyone Can Join”).

- IT receives an alert when employees add external users with "@competitor.com" in their email addresses to private groups or channels.
- IT receives an alert when an external user logs in after 30 days of inactivity.
- IT receives an alert via text message and email when an executive logs in from an unrecognized device and the location is not in New York City.
- IT receives an alert when someone from one region gains access to a group or distribution list in another region.
- IT receives an alert when any calendar is made public.
- IT receives an alert when any secondary calendar is created (e.g., a marketing calendar, an events calendar).
- IT receives an alert when automations run.

Measure user engagement.

This example is not related to security, but can be useful nonetheless. If users are reluctant to adopt a new SaaS application that was recently rolled out, IT can use Insights to monitor user engagement. If engagement is low, IT can create training materials. By encouraging adoption, IT helps users maximize performance and get more value out of SaaS applications.

- IT receives an alert when a user hasn't logged into an application in 30 days.

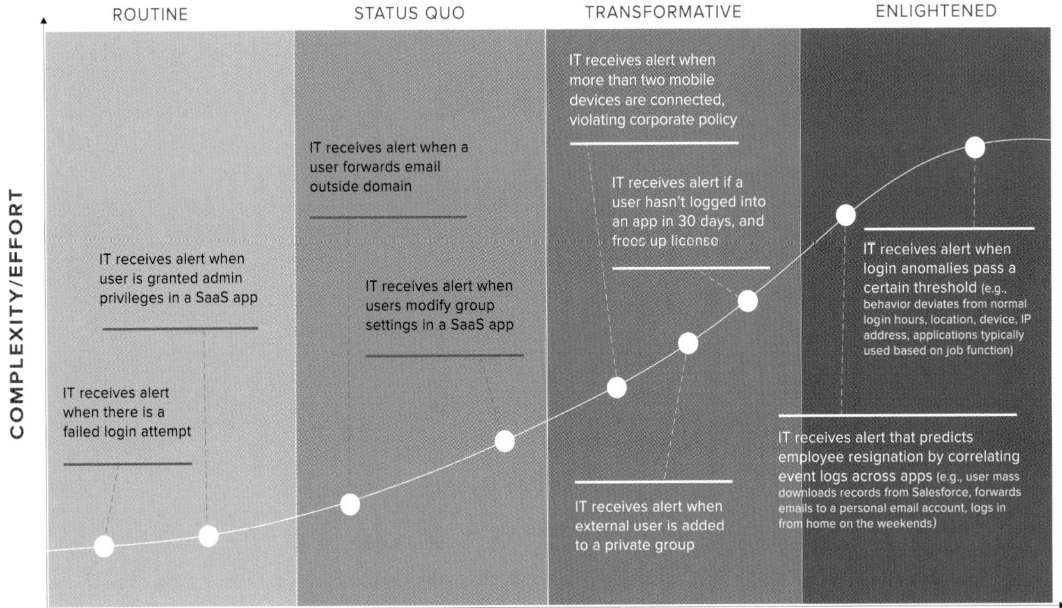

Figure 9

Use Cases: From Routine to Enlightened

IT can easily receive certain types of alerts in native admin consoles. Failed login attempts (or suspicious login attempts) are one example (in fact, this is probably *too* easy, as IT teams are often deluged with this type of alert. This diminishes the usefulness of this alert and creates alert fatigue because they are so common).

But alerts for user inactivity or for when an external user is added to a private group are more complex. These types of insights can be important for security, compliance, and even finance.

Insights into mass downloading or login anomalies across multiple SaaS applications, for example, are even more challenging to obtain. This would require manually examining event logs and correlating warning signs. Yet the insights they yield are also very useful, because they can provide predictive alerts and/or minimize security risks. ▲

Next: Action

Centralization, Discoverability, and Insights are about providing clarity. They collectively give IT a solid grasp of what data exists across an environment, how is that data is connected, and what critical events are happening in the environment.

The next two elements, Action and Automation, represent the other side of the equation. Once there is a clear picture of what data exists and which critical insights have surfaced, the next step is taking action against those data objects.

CHAPTER FOUR

ACTION

CHAPTER HIGHLIGHTS

ACTION

Action refers to the changes IT teams make to data objects to keep their environments running **securely, safely, and efficiently**.

IT is often hamstrung by **native SaaS application admin consoles**, which were not purpose-built for IT.

One of the biggest problems is that **IT cannot take action in bulk**, either within a single SaaS application or across multiple applications.

When organizations run multiple instances of SaaS applications, each instance is likely to have been set up differently with varying users, groups, security settings, etc. **Taking action becomes exponentially more cumbersome.**

Another major challenge around actions is that there's no searchable service catalog of all available SaaS admin actions. To discover all the various actions that can be taken on a data object, user, group, or third-party application in a SaaS application, **IT must pick through each admin console individually**.

To fully maximize this framework, **IT should ideally have the ability to audit and filter down a data set** and then take action against a single object (or a set of objects in bulk) that has been identified through a critical insight or alert.

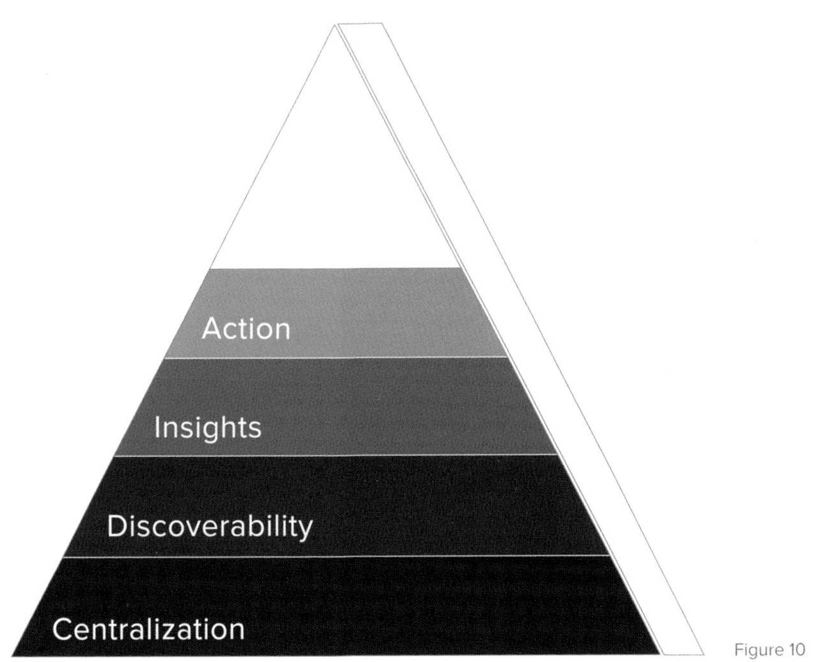

Figure 10

SaaS Application Management and Security Framework

ACTION

The Challenge: Carrying Out Operational Tasks Efficiently

"Action" refers to the routines and daily work that keeps an organization running. It's the changes IT teams make to users, groups, files, and data objects that keep the business running securely, safely, and efficiently. It's the activity IT teams engage in to remedy incidents. In essence, it's the constant pruning and maintenance of an environment.

Actions consist of administrative and operational functions. This includes creating objects (e.g., creating a user in a SaaS application), removing objects (e.g., deleting the

OAuth token for a risky third-party application), and changing a connection to another object (e.g., removing external collaborators from a file), among others.

Ensuring that day-to-day SaaS operations run smoothly (and securely) is no small feat. The number of tasks that need to be completed on a daily basis, the sheer volume of data objects across an organization, and the magnitude of SaaS sprawl are all tremendous. It's little wonder that SaaS-Powered Workplaces are twice as likely than average workplaces to say that managing users and assets is a challenge.[49] IT must manage users, memberships, calendars, groups, permissions, files, and more. Multiply this by dozens of applications and hundreds or thousands of users and you begin to understand why this is a major operational challenge.

50% of respondents do not agree or are unsure whether their organizations have the ability to manage and control user access to sensitive documents and how they are shared.

- Ponemon Institute

According to a survey of over 1,000 IT practitioners by the Ponemon Institute, 50% of respondents do not agree or are unsure whether their organizations have the ability to manage and control user access to sensitive documents and how they are shared.[50] While this is disturbing, it is not entirely surprising. IT administrators lack the correct tools to manage and control user access and sharing controls. They essentially lack Centralization, Discoverability, and Insights. Without these three elements, their hands are tied; they cannot take any action to protect sensitive data.

Why is this the case? For one, carrying out operational tasks efficiently is not easy. IT

---

49 "2017 State of the SaaS-Powered Workplace Report." BetterCloud, https://www.bettercloud.com/monitor/state-of-the-saas-powered-workplace-report/. Accessed 15 June 2017.

50 "Breaking Bad: The Risk of Unsecure File Sharing." Ponemon Institute, https://www.intralinks.com/platform-solutions/solutions/via/breaking-bad-risk-unsecure-file-sharing. Accessed 7 July 2017.

is often hamstrung by native SaaS application admin consoles that were not purpose-built for IT. They were built to help organizations work more effectively, which, to be sure, they do. But because they were not designed with IT in mind, they lack features that enable IT to work effectively and efficiently.

For instance, one of the most glaring problems is that IT cannot take bulk actions, either within a single SaaS application or across multiple applications. This feature, among others, does not exist natively in many admin consoles. Most admin consoles were built to take one-off actions (i.e., not tied to an ITIL process) and cover a subset of all available admin actions. As such, some actions are only available via APIs. Unless IT teams have experience in scripting, they have no way of knowing that these actions exist, impeding their ability to manage their environment efficiently. One example of such an action is removing all deleted users' future calendar events to free up calendar resources (e.g., conference rooms). IT teams must leverage APIs to delete these events one by one, and cannot delete them en masse.

Carrying out operational tasks efficiently is not easy. IT is often hamstrung by native SaaS application admin consoles, which were **not purpose-built for IT**.

The inability to take bulk action is also a large reason why many IT teams turn to scripting. Scripts are a cost-effective way to automate bulk operations that are otherwise time-consuming. However, scripting may not be a feasible long-term solution for many organizations. There are major security and compliance disadvantages—scripting lacks a traceable audit trail, as well as granular access roles—and when SaaS vendors change their APIs, even the most well-made scripts break. There are API quota limits as well. Turnover also poses a serious risk for organizations that rely on scheduled scripts that perform critical actions. If IT does not maintain up-to-date documentation or if a replacement hire lacks scripting knowledge, then there will inevitably be continuity issues. IT cannot gain central visibility by piecing together multiple scripts.

Additionally, when organizations run multiple instances of SaaS applications, management and operational security become even more complex. Data is fragmented across different instances of the same application, so taking action is much more difficult. Instead of logging into one native admin console for a single application, admins must log into multiple consoles to manage all the disparate instances of the same application. Each instance is likely set up differently with varying users, groups, security settings, etc. Taking action becomes exponentially more cumbersome.

Another major challenge related to actions is that there's no searchable service catalog of all available SaaS admin actions. To discover all the various actions that can be taken on a data object, user, group, or third-party application in a SaaS application, IT must pick through each admin console individually. There is no central hub. This is exactly why Centralization and Discoverability are so critical: if IT can't see all of its SaaS data in one central place, then it cannot take action against data objects in an effective manner.

The Action Landscape

As mentioned above, there are alternative solutions (like scripting) that allow IT teams to circumvent the limitations of native admin consoles. However, while these are viable solutions for some organizations, they do come with drawbacks. Here are a few advantages and disadvantages:

VENDOR CATEGORY	ACTION FUNCTIONALITY	ADVANTAGES	DISADVANTAGES
Scripts	Can address every API action that is exposed by SaaS vendors	Custom and granular way to address remediation actions	Cumbersome to develop Tough to maintain with updates and new releases of APIs Some require the Client ID and OAuth Secret to be un-encrypted Not captured in audit logs (lack of auditability) Require super admin privileges to run
Native SaaS application admin consoles	Can provide a way to let IT take specific actions inside the application	Natively supported and maintained by the vendor Intuitive to use	Not comprehensive; always addresses the lowest common denominator of functions Only works for a single application and does not scale to include multiple instances of the same application
SaaS Application Management and Security Platform	A rich catalog of SaaS application admin actions is available in such platforms, allowing for deep stateful app awareness at the user and data object level through extensive API integrations	Aggregated catalog of almost any SaaS application admin config or task action is possible User and data awareness allows IT to enact instant one-off or bulk actions without the need for scripts or multiple steps	Reliant on public APIs that are exposed by each SaaS application platform for admin config and task actions

The Solution: A Centralized, Contextual, Searchable Action Engine

Admin consoles must provide functionalities that facilitate SaaS management and operations. To manage a SaaS environment effectively, IT needs to be able to take action against data objects in a way that is fast, efficient, and secure. This is especially true for situations where IT performs the same actions in bulk across SaaS applications. SaaS administrators often need to take similar actions against similar objects across multiple SaaS applications. For example, IT might want to take the same action (e.g., resetting passwords) for similar users (e.g., everyone in the New York office) every three months.

To fully maximize this framework, IT should ideally have the ability to audit and filter down a data set and then take action against a single object (or a set of objects in bulk) that has been identified through a surfaced critical insight or alert. At that point, IT can leverage insights to take action to remediate the problem.

Taking action against a set of data objects requires write scopes and interaction with external APIs in bulk. For instance, if a SaaS administrator wants to edit permissions for 1,000 groups, he must make 1,000 (or more) separate API calls to change the properties for each of those 1,000 groups and fulfill all those requests in a job queue. While it is possible for an IT team to build an application to make these calls, it is far from simple. Mass processes like these require ample domain knowledge vis-a-vis interacting with APIs in bulk.

Additionally, every provider has different limits and quotas. Some SaaS providers have daily API quotas, while others have queries per second (QPS) limits. Furthermore, APIs change often and may break applications unexpectedly. For example, if v1 of a SaaS provider's activity feed is deprecated, SaaS administrators must rewrite their application in order to listen to v2 of the new activity feed. Broken applications pose a serious operational problem if IT is using them to carry out critical day-to-day actions in a SaaS environment. Keeping abreast of API changes would require monitoring internal logs for errors of API failures, carrying out automated testing, and more.

Use Cases: How Action Enables SaaS Management Success

These sample use cases below illustrate how the ability to take bulk action saves time and increases efficiency:

- ▲ IT resets all users' passwords across all its SaaS applications with one click.
- ▲ IT bulk updates the mailing address, phone number, and email signatures across all SaaS applications when an organization undergoes a rebrand, acquisition, merger, or divestiture.
- ▲ IT unshares any publicly shared files (in bulk) that contain the word "Confidential" and are owned by any user in finance.
- ▲ Multiple employees move internally to a different team in a different office location, and IT changes their group memberships (in bulk) to mirror an existing team member's.
- ▲ IT creates multiple groups and calendars and modifies their settings in bulk.
- ▲ IT changes ownership of files/permission levels in bulk for an entire business unit.

The following examples build upon the use cases mentioned in the previous section of this framework. The action directly remediates a problem that has surfaced through an insight.

INSIGHT	ACTION
IT receives an alert when a user mass downloads data from Dropbox and Salesforce within seven days, forwards emails to a personal email account, and logs in repeatedly from home on the weekends.	IT suspends user, sends a Slack message to the user's manager, and opens a ticket in the ITSM with the security team.
IT receives an alert when any user in the finance department shares files publicly.	IT removes public sharing and changes the files back to private.
IT receives an alert when any user in the finance department sets up an email forwarding policy and turns off two-factor authentication.	IT suspends user.
IT receives an alert when a user installs a third-party application that hasn't been reviewed yet and violates security criteria.	IT revokes access to third-party application.
IT receives an alert when a new super admin is added to an application.	IT revokes access or creates new access role.
IT receives an alert when an external user has been inactive for 30 days.	IT disables inactive external user's accounts **across all SaaS applications.**

INSIGHT	ACTION
IT receives an alert when a user modifies group settings and changes them to public (or "Anyone Can Join").	IT changes sharing options to a more private setting.
IT receives an alert when employees add external users with "@competitor.com" in their email addresses to private groups or channels.	IT removes external users with "@competitor.com" in their email addresses from all groups **across SaaS applications.**
IT receives an alert when any file containing the word "Confidential" is shared publicly.	IT un-shares all files containing the word "Confidential" **across all SaaS applications**.
IT receives an alert when a user hasn't logged into an application in 30 days.	(Option #1): The user isn't using the application because they're unfamiliar with how it works, so IT sends them an email with helpful training materials. (Option #2): The user isn't using the application because they no longer need it, so IT frees up the license and assigns it to someone else.
IT receives an alert via text message and email when an executive logs in from an unrecognized device and the location is not in New York City.	IT notifies executive (or executive's assistant) via Slack or SMS about the unusual activity and gives them a chance to confirm before resetting the password.

INSIGHT	ACTION
IT receives an alert when someone from one region gains access to a group or distribution list in another region.	IT revokes access for user.
IT receives an alert when any calendar is made public.	IT modifies settings back to private.
IT receives an alert when any secondary calendar is created (e.g., a marketing calendar, an events calendar).	IT modifies settings back to private.
IT receives an alert when a user's account has been deleted.	IT deletes future recurring events on departed employee's calendar to free up calendar resources (e.g., conference rooms).
IT receives an alert when new sales hires do not belong to any groups/channels.	IT adds new sales hires to all relevant sales groups/channels **across all SaaS applications.**

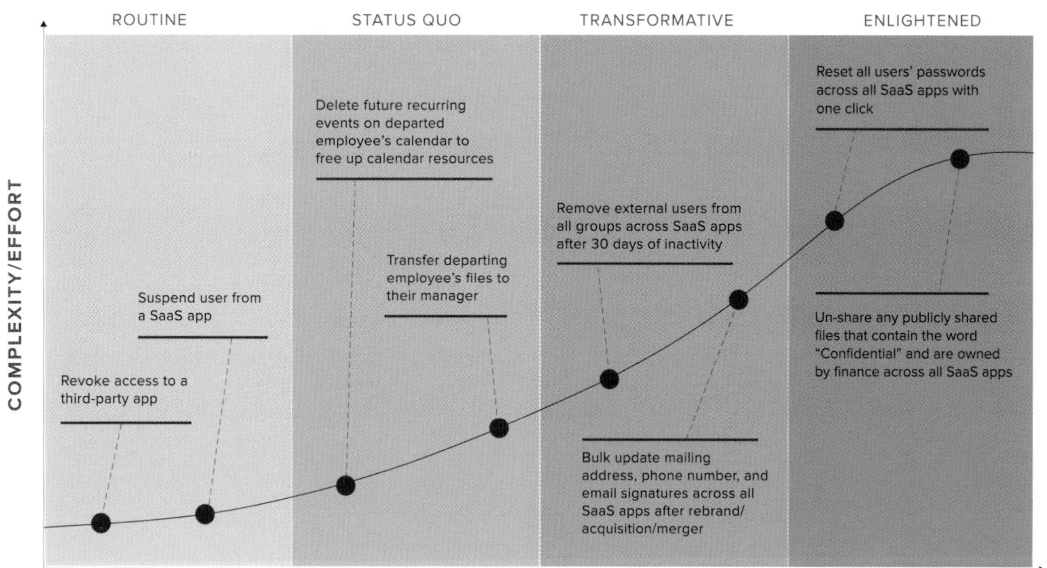

Figure 11

Use Cases: From Routine to Enlightened

Figure 11 depicts several Action use cases. Examples of very simple, routine use cases include revoking access to a single third-party application and suspending a user from a SaaS application. Both can be executed in a few steps inside a native admin console. However, note that these are one-off actions (i.e., not taken in bulk). Doing these actions one by one is time-consuming for IT and brings little value.

However, other use cases along the continuum do become more valuable for IT. Removing external users from all groups across SaaS applications after 30 days of inactivity, for example, is immensely useful. It ensures that no contractor has excessive access to corporate data past their contract end date. This protects sensitive data and reduces the risk of privileged access abuse. However, this task is also much more complex for IT, as it requires multiple steps in multiple SaaS applications and likely hours of manual work.

And, finally, there are use cases such as “un-share any publicly shared files that contain the word ‘Confidential’ and are owned by finance across all SaaS applications.” Un-sharing these files would first require Discoverability (finding files that are publicly shared and contain the word ‘Confidential’ *and* are owned by finance users). IT cannot easily get Discoverability, but actions like this would greatly help secure an environment.

Or consider another example use case: resetting all users’ passwords across all their SaaS applications with one click. IT typically has to log into multiple admin consoles and reset passwords separately, app by app. Doing it in bulk across applications for all users would save IT immense amounts of time. These kinds of actions are what IT needs in order to keep an environment secure and maximize efficiency. ▲

Up Next: Automation

Taking action against data objects is only half the battle. SaaS administrators need the next element of the framework to automate all of these actions in order to secure their environment and work efficiently.

CHAPTER FIVE

AUTOMATION

CHAPTER HIGHLIGHTS
AUTOMATION

As companies grow and adopt more SaaS applications, on- and offboarding processes and user lifecycle management become exponentially more **repetitive, complex, and time-consuming.**

Not only does the volume of work increase, but so does the pace. **Companies are reaching a breaking point.** If it hasn't already, IT will hit a wall where it can no longer keep up with the tedious, repetitive work that SaaS applications create.

Actions taken in admin consoles are manual and often error-prone, which **increases the risk of a security incident.**

Automation isn't just "nice to have" in SaaS environments; it's necessary to manage the increasing volume of work.

The key to any valuable IT automation, much like alerts, is the ability to be highly granular and customizable based on certain attributes. They should also naturally be enabled once the same pattern occurs repeatedly (i.e., when this occurs, if these conditions are true, then take these actions). This way, when specific triggers take place, **the problem will automatically be remediated, thereby creating a self-healing environment.**

Additionally, because **it can automatically enforce compliance**, automation is critical in SaaS environment when it comes to operational security.

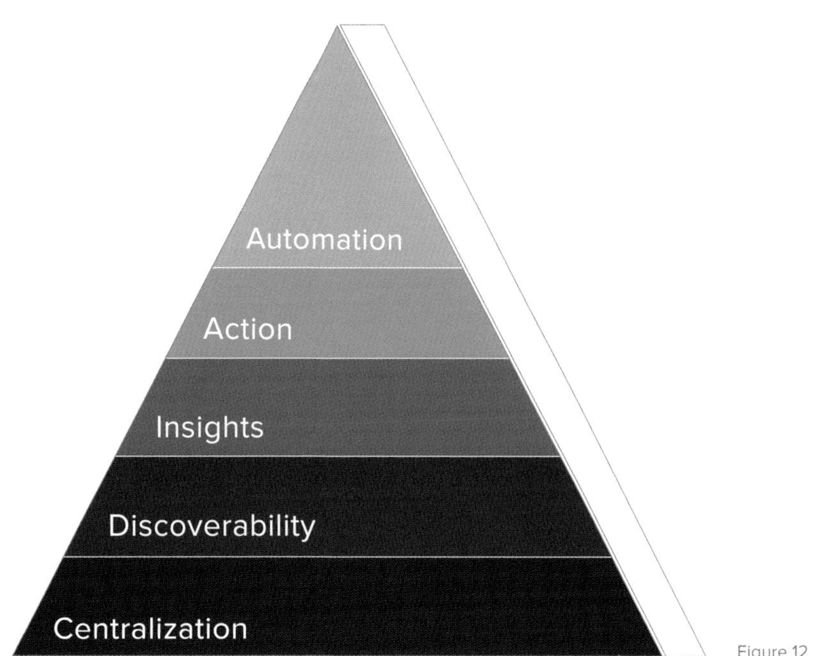

Figure 12

SaaS Application Management and Security Framework

AUTOMATION

Challenge #1: SaaS Adoption Increases Manual, Repetitive Work

Managing and operating SaaS environments is, and will continue to be, detrimental to time management. As companies grow and adopt more SaaS applications, on- and offboarding processes and user lifecycle management become exponentially more repetitive, complex, and time-consuming.

For instance, onboarding and offboarding processes consist of multiple repetitive steps. On average, it takes 10 steps to offboard a user from a single SaaS application.

Some applications require as many as 25. That number rises in multi-SaaS environments as IT must multiply those steps across dozens of applications. In fact, SaaS-Powered Workplaces are 2.5 times more likely to say that automating repetitive tasks is a challenge.[51]

Managing user lifecycle changes also becomes more complex. For example, updating group memberships correctly in one SaaS application is one thing; doing it accurately across multiple applications is quite another. Since each application is its own island—siloed and independent of others—cross-app work is full of friction. There's no easy way to repeat actions across applications. IT must log into dozens of admin consoles individually to take actions, which becomes cumbersome and complex.

Research shows that **41% of IT professionals** say they either believe former employees still have access to company data or they don't know, and 53% say they often or very often fail to give new hires access to the right applications.

Furthermore, not only does the volume of work increase, but so does the pace. Departing employees need to be offboarded quickly and completely lest they retain access to sensitive data after they leave the company. Likewise, new employees expect to be onboarded properly and given the appropriate access to applications so they can hit the ground running on day one. Yet IT cannot keep up with this pace: research shows that 41% of IT professionals say they either believe former employees still have access to company data or they don't know, and 53% say they often or very often fail to give new hires access to the right applications.[52]

---

51 "2017 State of the SaaS-Powered Workplace Report." BetterCloud, https://www.bettercloud.com/monitor/state-of-the-saas-powered-workplace-report/. Accessed 15 June 2017.

52 "Trends in Cloud IT." BetterCloud, https://www.bettercloud.com/monitor/research. Accessed June 15 2017.

Companies are reaching a breaking point. A recent report found that nearly 80% of business leaders said that data from mobile devices and the Internet of Things was accelerating the pace of work, and 91% agree that skilled employees spend too much time on administrative tasks.[53]

IT has three options today:

1. Scale IT staff alongside SaaS application usage and adoption

2. Block additional SaaS usage and adoption

3. Maintain the status quo and run the risk of offboarding employees improperly. This means employees may still retain access to critical data after they leave, and/or the organization may incur extraneous costs for unused licenses, etc.

For most organizations, none of these options are acceptable, leaving IT between a rock and a hard place. Ultimately, if it hasn't already, IT will hit a wall where it can no longer keep up with the tedious, repetitive work that SaaS applications create.

Challenge #2: Human Error Means Mistakes Will Always Be Made

SaaS applications have been built with an emphasis on openness, accessibility, collaboration, and ease of use, but with these qualities come chances for user error. Actions taken in admin consoles are manual and often error-prone and thus increase the risk of a security incident.

To err is human. But Amazon's massive cloud service outage in early 2017 is a reminder that human error, such as a single typo, can have disastrous consequences (according to one estimate, Amazon's outage cost S&P 500 companies $150 million

53 "State of Work 2017 Report." ServiceNow, https://www.servicenow.com/content/dam/servicenow/documents/whitepapers/sn-state-of-work-report-2017.pdf. Accessed June 15 2017.

and US financial service companies $160 million in lost revenue).[54]

Additionally, repetitive tasks can easily lead to mistakes. In a study reported by *The Economist*, researchers gave volunteers a "flanker" test, which measured performance in the presence of distracting information. "[Volunteers] were asked to respond as quickly as possible to the direction of an arrow flanked by other arrows that point in the same or opposite direction. Although the task is simple and repetitive, to keep providing the right answer demands a fair bit of brain power: people make a mistake about 10% of the time."[55]

This makes sense. When repetitive tasks are performed manually, there will always be some percentage of human error. Humans are fallible, after all. Research shows that during "monotonous, simple, and repetitive operational and technical work, one mistake might happen in every 100-1,000 operations."[56] An error rate of 1% or 0.1% may not sound alarming, but in industries like aviation (which this statistic is referring to), mistakes can have fatal consequences. *Any* amount of human error is unacceptable.

Amazon's massive cloud service outage in early 2017 is a reminder that human error, such as a single typo, **can have disastrous consequences**.

Human error is an ever-present danger. According to Boeing, approximately 80% of airplane accidents are due to human error[57] (pilots, air traffic controllers, mechanics, etc.—consider the repetitive nature of cockpit procedures, aircraft maintenance, and so forth). Similarly, according to the Accident Research Team at Volvo Trucks, human error is involved in as many as 90% of all truck accidents.[58]

54 "One Amazon Employee's 'Human Error' May Have Cost the Economy Millions." Vanity Fair, http://www.vanityfair.com/news/2017/03/one-amazon-employees-human-error-may-have-cost-the-economy-millions. Accessed June 15 2017.

55 "It Is Possible to Predict Human Errors from Brain Activity." The Economist, http://www.economist.com/node/11088585. Accessed 15 June 2017.

56 Wang, Jinsong, Proceedings of the First Symposium on Aviation Maintenance and Management - Volume II (Heidelberg: Springer, 2014), page 542. https://books.google.com/books?id=alXFBAAAQBAJ&pg=PA542#v=onepage&q&f=false.

57 Rankin, William. "MEDA Investigation Process." Boeing, http://www.boeing.com/commercial/aeromagazine/articles/qtr_2_07/AERO_Q207_article3.pdf. Accessed 17 July 2017.

58 "European Accident Research and Safety Report 2013." Volvo, http://www.volvotrucks.com/SiteCollectionDocuments/VTC/Corporate/Values/ART%20Report%202013.pdf. Accessed 17 July 2017.

Additionally, research shows that medical errors may be the third leading cause of death in the US, right behind heart disease and cancer; doctors estimate there are at least 251,454 deaths due to medical errors annually in the United States.[59]

Because much of the work in IT is manual and repetitive, there is ample room for human error to creep in. Of course, human error in IT (probably) won't cause any fatalities, but it can still cause disastrous outages, security catastrophes, irreparable data loss, and more. Without the aid of automation, IT can succumb to slips (e.g., making a typo and causing an outage), lapses (e.g., getting distracted and forgetting a critical security step in an offboarding procedure), errors in judgment (e.g., making a mistake because of improper training or alert fatigue), or attempting workarounds to save time (e.g., taking a well-meaning shortcut that fails). Additionally, according to *Infosecurity Magazine*, human error is the number one cause of data breaches.[60]

Challenge #3: Ex-Employees Can Retain Access If Offboarding Processes Are Carried Out Incorrectly

Automation can be immensely beneficial to manual processes like offboarding. Organizations cannot afford to make mistakes here—if an IT team forgets a step and offboarding is incomplete, former employees can pose huge security threats.

A survey by Osterman Research revealed that 89% of ex-employees retained access to Salesforce, PayPal, email, SharePoint, Facebook, and other sensitive corporate applications.[61] What's worse, 45% retained access to "confidential" or "highly confidential" data, and 49% logged into an account after leaving the company. The ramifications are endless: compliance failures, stolen intellectual property, data breaches, data loss, etc. According to the report, these figures are not surprising,

59 Christensen, Jen and Elizabeth Cohen. "Medical errors may be third leading cause of death in the U.S." CNN, http://www.cnn.com/2016/05/03/health/medical-error-a-leading-cause-of-death/index.html. Accessed 17 July 2017.
60 "It Shouldn't Matter how Many USBs are Lost." Infosecurity Magazine, https://www.infosecurity-magazine.com/blogs/it-shouldnt-matter-how-many-usbs/. Accessed 15 June 2017.
61 "The Ex-Employee Menace: Intermedia's 2014 SMB Rogue Access Report." Osterman Research and Intermedia, https://www.intermedia.net/Reports/RogueAccess. Accessed 19 July 2017.

as most offboarding processes are disorganized and incomplete.

45% of ex-employees retained access to "confidential" or "highly confidential" data, and 49% logged into an account after leaving the company.

- Osterman Research

In fact, rogue access is such a significant problem that it even impelled the FBI to issue a public warning about it, stating that: "The exploitation of business networks and servers by disgruntled and/or former employees has resulted in several significant FBI investigations in which individuals used their access to destroy data, steal proprietary software, obtain customer information, purchase unauthorized goods and services using customer accounts, and gain a competitive edge at a new company [...] A review of recent FBI cyber investigations revealed victim businesses incur significant costs ranging from $5,000 to $3 million due to cyber incidents involving disgruntled or former employees."[62]

If a critical process like offboarding is done manually, it leaves IT vulnerable to mistakes. Automation helps reduce (or altogether eliminate) errors and thereby minimize security risk.

The Automation Landscape

It is no surprise that many automation products exist on the market today. After all, automation improves productivity and reduces human error. However, many automation vendors fall short. Here are a few benefits and challenges:

62 "Public Service Announcement: Increase in Insider Threat Cases Highlight Significant Risks to Business Networks and Proprietary Information." The Department of Homeland Security, https://www.ic3.gov/media/2014/140923.aspx. Accessed 19 July 2017.

VENDOR CATEGORY	AUTOMATION FUNCTIONALITY	ADVANTAGES	DISADVANTAGES
Scripts	Can address every API action that is exposed by SaaS vendors	Custom and granular way to address remediation actions	Cumbersome to develop Tough to maintain with updates and new releases of APIs Some require the Client ID and OAuth Secret to be un-encrypted Not captured in audit logs (lack of auditability) Require super admin privileges to run
BPM workflow automation tools (Zapier, Azuqua, etc.)	Extensive catalog of tasks and actions that can be automated for applications	Flexibility and large variety of actions provided across a large set of applications Best suited for handling business process workflow automation	No statefulness or context (user or data object) available for actions. Such platforms are not suitable for IT configuration and task automation
IT Service Management (ITSM)/IT Operations Management/ (ITOM)	Automation is limited to the integrations available in the app stores and marketplaces of such platforms	Simple user account provisioning automation is available with third-party integrations	No native awareness of SaaS applications as "first-class" IT assets within such platforms, which makes it impossible for orchestration or automation of any actions inside SaaS apps The "action service catalog" of SaaS applications is a big gap. The closest functionality available is to orchestrate actions for IaaS workloads
SaaS Application Management and Security Platform	A rich catalog of SaaS application admin actions is available in such platforms, allowing for deep stateful application awareness at the user and data object level through extensive API integrations	Automation of almost any SaaS application admin config or task action is possible, along with orchestration and security policy enforcement across applications Provides flexibility to orchestrate actions through existing EMM and IDaaS platforms with full application statefulness and user and data context	Reliant on public APIs to be exposed by each SaaS app platform for admin config and task actions The degree of automation for an app is restricted to the API capabilities of the SaaS application

The Solution: Automation

As the pinnacle of this framework, Automation is the most powerful element. It's the *sine qua non* of a high-performing IT operation.

Automation isn't just "nice to have" in SaaS environments; it's necessary to manage the increasing volume of work. A recent report found that 86% of executives said that they would need greater automation by 2020, and nearly half (46%) said that the breaking point was coming by 2018.[63] IT will soon have no choice but to automate manual, repetitive work. Gartner estimates that in 2017, 75% of enterprises will have more than four diverse automation technologies within their IT management portfolios, up from less than 20% in 2014.[64]

> A recent report found that **86% of executives** said that they would need greater automation by 2020—and nearly half (46%) said that the breaking point was coming by 2018.
>
> - ServiceNow

The key to any valuable IT automation, much like alerts, is the ability to be highly granular and customizable based on certain attributes. "One size fits all" automations are rarely beneficial. For example, an onboarding process might look the same for many users in an organization, but factors like their title, location, or department will cause variances.

The other key to making automation powerful for SaaS administrators is that it should naturally be "enabled" once the same pattern occurs repeatedly (i.e., when this occurs, if these conditions are true, then take these actions). This way, when specific

63 "State of Work 2017 Report." ServiceNow, https://www.servicenow.com/content/dam/servicenow/documents/whitepapers/sn-state-of-work-report-2017.pdf. Accessed 15 June 2017.

64 "Gartner Predicts." Gartner, http://www.gartner.com/binaries/content/assets/events/keywords/data-center/dci5/gartner-predicts-for-it-infrastructure-and-operations.pdf. Accessed 15 June 2017.

triggers take place, the problem will automatically be remediated, thereby creating a self-healing environment.

In this framework, Automation is the culmination of all the elements that have come before it—having all of the previous elements of the framework in place therefore is critical. The only way IT can automate effectively is by building upon every single one of the previous elements in this framework.

When specific triggers take place, the problem will automatically be remediated, thereby **creating a self-healing environment.**

Automation relies on Centralization and Discoverability to view and filter millions of SaaS provider events and figure out what data has changed. It then relies on Insights to surface any critical events or correlations, listening for the ones that matter. It then relies on Action to coordinate multiple external API calls and carry out those tasks. And finally, Automation carries out all those critical actions based on triggers, automatically stringing together disparate data objects in choreographed workflows. True SaaS automation automates granular actions from multiple areas across SaaS lifecycle management, including memberships, settings, third-party applications, and more. It provides IT deep control.

Since IT is responsible for keeping their organizations compliant with internal policies and industry regulations like SOX, PCI, HIPAA, and FERPA, Automation ensures nothing falls through the cracks and that sensitive information remains confidential. For example, if IT automatically receives an alert that a file is being shared publicly, that by itself is not tremendously useful. But if IT receives an alert based on specific triggers, and then the file owner automatically receives a notification email and the file's sharing setting is automatically changed to Private—all in one workflow—then that becomes much more valuable. Automating one action reduces the chance of human error, but automating multiple related actions reduces it even further.

Furthermore, not only does Automation reduce (or remove) human error and improve consistency, it also allows IT to move up the value chain in an organization. IT teams must decipher noisy alerts and manually complete byzantine on- and offboarding processes all day. This leaves them with little time to work on higher-value, long-term projects. Examples of such strategic work include internal technology consulting, selecting new applications, etc. Automation frees up IT. It allows them to be more productive. It enables IT professionals to serve as strategic advisors, focus on new business solutions, and deliver business results.

"CIOs should aim to find a third-party solution that can simplify the management of their IT by absolving humans of repetitive tasks, and then channeling the brainpower of IT staff into tasks that require problem solving, creativity, ingenuity or innovation," writes *Wired*. "With the reduction of human error, we might even see IT get smarter still. As engineers reduce the amount of time spent putting out their own fires, they will instead have the time and resources to improve overall performance and offerings, allowing IT to expand far beyond its current limits."[65]

But IT cannot evolve its role and become an agent of change if it is mired in tedium every day. Automation enables innovation, shifting IT's job focus from utility to digital transformation. In fact, IT *must* evolve its role to stay competitive: Gartner predicts that by 2020, 50% of the CIOs that have not transformed their capabilities will be displaced from the digital leadership team.[66]

Moreover, automation improves talent retention. "Are you looking to hire the one-year, Tier 1 helpdesk burnout that creates and offboards accounts?" says Dave Jackson, Director of Infrastructure at WeWork. "Or will you use the headcount for a more skilled support engineer to manage that 30,000-foot view of automation (e.g., account workflows, MDM, etc.)?"

---

65 Crane, Jonathan. "Eradicate Human Error Without Limiting IT." Wired, https://www.wired.com/insights/2012/11/eradicating-human-error-without-limiting-it/. Accessed 17 July 2017.
66 "Gartner Predicts." Gartner, http://www.gartner.com/binaries/content/assets/events/keywords/data-center/dci5/gartner-predicts-for-it-infrastructure-and-operations.pdf. Accessed 15 June 2017.

Use Cases: How Automation Enables SaaS Management Success

It is hard to overstate how critical and powerful Automation is in SaaS environments. From a compliance perspective, Automation is invaluable. Currently, the only way for many IT professionals to ensure policies are enforced is through manual reviews of logs and data in SaaS applications. IT can avoid this manual work and instead automate policy enforcement, saving time and maintaining the operational security of their environment.

The following are examples of use cases illustrating how Automation can enable success in different types of situations:

Automate the onboarding process.
Automation can help ensure that new hires have the right access to the right data when they join a company. This creates a smoother onboarding process, allowing new hires to hit the ground running.

- ▲ When a new user belongs to the sales team (or has Person X as a manager, or has "Senior Sales" in his title), automatically grant them a role in a CRM; automatically grant them access to email, e-signature, and file management programs; automatically grant them access to the appropriate folders; automatically add them to the appropriate groups; and automatically send a reminder message to their manager to schedule a one-on-one meeting.

Automate user lifecycle management.

- ▲ When a user changes teams, automatically move them to the correct groups; grant them appropriate access to any files, folders, and/or new applications required for their new role; revoke access to the groups, files, folders, applications, etc. that they shouldn't have access to anymore; and update their email signature.

- Automatically place users into groups based on their business unit, location, subsidiary company, employee status, country, or manager. When any of this information changes, automatically update their group membership(s).

Automate the offboarding process.

Automation can also ensure that the offboarding process is seamless. If offboarding is manual, steps can easily be overlooked, resulting in data loss, retention of access rights, and so forth. Automation can also help coordinate efforts across teams such as IT and HR.

- When a user is deactivated in an HRIS, automatically suspend or delete the same user in other SaaS applications.
- When a user is suspended, automatically hide the user in the directory, reset the user's password, add an auto-reply message, delegate email access to their manager, transfer ownership of sites, calendars, files, and groups to their manager, and revoke access to third-party applications.

Automatically flag suspicious user behavior and remediate issues.

- When a user downloads more than 100 accounts in Salesforce within seven days, automatically suspend their account, send a Slack message to their supervisor, and open a ticket with the security team.
- Based on suspicious login behavior, automatically lock down a user's account (e.g., suspend user, reset passwords across SaaS applications, reset sign-in cookies, etc.).

Automate calendar resource selection when users create meetings.

- Automatically assign calendar resources (e.g., conference rooms) based on the meeting title, number of people, and attendees to maximize usage efficiency.

The examples below build upon the use cases mentioned in the previous chapters of this book. Automation automatically takes action to remediate a problem that has been surfaced through an insight. This increases the speed of remediation, ensuring that the problem never becomes a full-blown security issue. By automating alert resolution, IT can also essentially automate compliance and ensure that no sensitive data is inadvertently exposed.

INSIGHT	AUTOMATION
IT receives an alert when a user mass downloads data from Dropbox and Salesforce within seven days, forwards emails to a personal email account, and logs in repeatedly from home on the weekends.	IT **automatically** suspends user, sends a Slack message to the user's manager, and opens a ticket in the ITSM with the security team.
IT receives an alert when any user in the finance department shares files publicly.	IT **automatically** removes public sharing and changes the files back to private.
IT receives an alert when any user in the finance department sets up an email-forwarding policy and turns off two-factor authentication.	IT **automatically** suspends user.
IT receives an alert when a user installs a third-party application that hasn't been reviewed yet and violates security criteria.	IT **automatically** revokes access to third-party application.
IT receives an alert when a new super admin is added to an application.	IT **automatically** revokes access or **automatically** creates new access role.

INSIGHT	AUTOMATION
IT receives an alert when an external user has been inactive for 30 days.	IT **automatically** disables inactive external user's accounts across all SaaS applications.
IT receives an alert when a user modifies group settings and changes them to public (or "Anyone Can Join").	IT **automatically** changes sharing options to a more private setting.
IT receives an alert when employees add external users with "@competitor.com" in their email addresses to groups or channels.	IT **automatically** removes external users with "@competitor.com" in their email addresses from all groups across SaaS applications.
IT receives an alert when any file containing the word "Confidential" is shared publicly.	IT **automatically** un-shares all files containing the word "Confidential" across all SaaS applications.
IT receives an alert when a user hasn't logged into an application in 30 days.	(Option #1): The user isn't using the application because they're unfamiliar with it how it works, so IT **automatically** sends them an email with helpful training materials. (Option #2): The user isn't using the application because they no longer need it, so IT **automatically** frees up the license and assigns it to someone else.

INSIGHT	AUTOMATION
IT receives an alert via text message and email when an executive logs in from an unrecognized device and the location is not in New York City.	IT **automatically** notifies executive (or executive's assistant) via Slack or SMS about the unusual activity and gives them a chance to confirm before resetting the password.
IT receives an alert when someone from one region gains access to a group or distribution list in another region.	IT **automatically** revokes access for the user.
IT receives an alert when any calendar is made public.	IT **automatically** modifies settings back to private.
IT receives an alert when any secondary calendar is created (e.g., a marketing calendar, an events calendar).	IT **automatically** modifies settings back to private.
IT receives an alert when a user's account has been deleted.	IT **automatically** deletes future recurring events on the departed employee's calendar to free up calendar resources (e.g., conference rooms).
IT receives an alert when new sales hires do not belong to any groups/channels.	IT **automatically** adds new sales hires to all relevant sales groups/ channels across all SaaS applications.

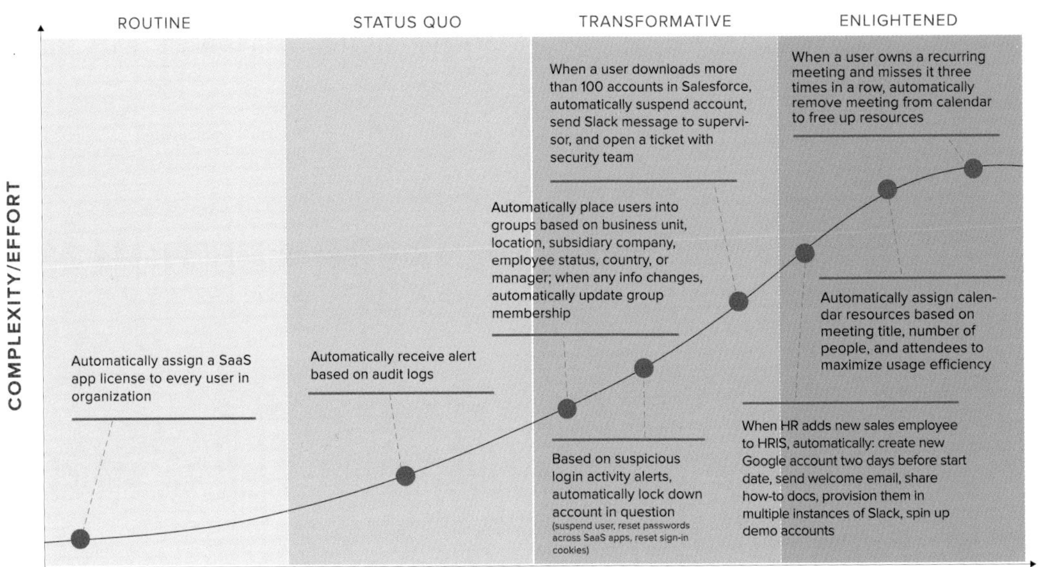

Figure 13

Use Cases: From Routine to Enlightened

An easy task to perform natively is to automatically assign a SaaS application license to every user in an organization. This is simple but the impact is minimal. It's a routine task.

But the apex for IT in terms of Automation are the use case examples in the "Enlightened" stage. These are tasks that would save IT hours of manual work and prevent dangerous human error. One example is an entirely automated "zero-touch" onboarding process that would take place once HR added a new employee to an HRIS. A new Google account would automatically be created two days before the user's start date. The user would automatically receive a welcome email containing helpful how-to guides. The provisioning would be dynamic; if the user was in sales, they would automatically be granted access to the right groups, files, and folders, as well as multiple instances of Slack. Demo accounts would be automatically spun up for them as well. These automated onboarding processes would essentially "image" a user and be tremendously valuable to IT. ▲

Up Next: Delegation and Auditability

All of these five elements work together, but Delegation and Auditability hold the entire framework together. This refers to the ability to audit activity and delegate administrative privileges to others in the IT department and beyond (e.g., HR or security teams), so that they can execute on these operational tasks.

CHAPTER SIX

DELEGATION & AUDITABILITY

CHAPTER HIGHLIGHTS

DELEGATION & AUDITABILITY

IT must implement the principle of least privilege. Having too many super admins is an inherent security risk. Every additional administrator causes a **linear-to-exponential growth in risk.**

However, this is impossible to do for two reasons. First, IT can't view its team's admin privileges across applications in one place—there's no unified view. Second, many of the less mature SaaS applications offer only binary options: super admin or end user, with nothing in between. **IT teams may not necessarily want to grant an employee *carte blanche*, but they are left with no other option**.

To fix this problem**, IT must have the ability to create custom CRUD roles** with specific, granular privileges for the data objects and controls across SaaS applications. IT can then delegate limited privileges to different business units and employees.

Auditing is also a major blind spot for IT. There is no easy way to thoroughly track an admin's actions in SaaS applications. An IT administrator would have to download logs from each SaaS application and parse through them one by one, manually correlating events across all of them.

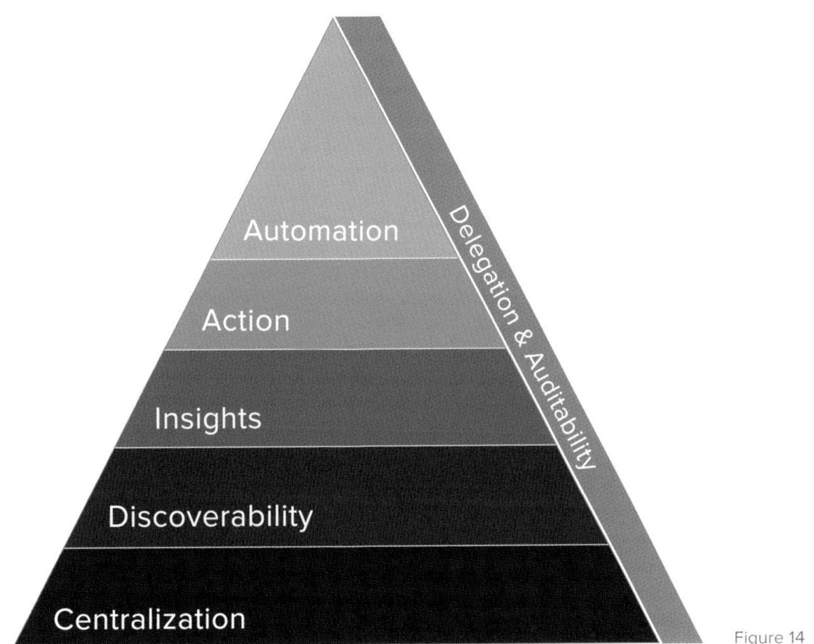

Figure 14

SaaS Application Management and Security Framework

DELEGATION AND AUDITABILITY

Delegation and Auditability aptly enclose the entire framework and hold it together.

DELEGATION

Challenge #1: Granting Employees the Ideal Level of Access

In their 2016 Data Breach Investigations Report, Verizon writes, "Be careful who you give privileges to and to what degree. It makes sense to give the valet attendant your keys to park your car, but not to hand over your credit cards as well."[67]

---

67 "2016 Data Breach Investigations Report." Verizon, http://www.verizonenterprise.com/verizon-insights-lab/dbir/2017/. Accessed 27 February 2017.

Indeed, it's critical to grant employees the right level of SaaS administrative privileges: not too much, not too little, and for the shortest time needed.

On one hand, having too many super admins is an inherent security risk. "Every additional administrator causes linear-to-exponential growth in risk," according to InfoWorld.[68]

Even if an IT director has the world's most trustworthy people on his team, each extra admin is a security threat. It's not a matter of integrity (or lack thereof); it's an objective security matter. "Every additional admin doesn't just increase his or her own risk; if they're compromised, they add to the takedown risk of all the others. Each admin may belong to groups others do not. If a hacker compromises A and gets to B, B may more easily lead to C, and so on," writes InfoWorld.[69]

> "Every additional administrator causes **linear-to-exponential growth** in risk."
>
> - InfoWorld

Accidental misuse of one's privileges, to say nothing of deliberate, malicious abuse, is a glaring security threat. Privileged user abuse can have severe ramifications. For example, an engineer with privileged user access at American Superconductor, a US-based global energy company, was enticed by a Chinese company to steal source code and other intellectual property from his employer in 2011. As a result of his theft, the company lost three quarters of its revenue, half of its workforce, and more than $1 billion in market value.[70]

But on the other hand, giving employees insufficient privileges is a hindrance to productivity. It means a small group of people are responsible for all of the day-to-day management. It slows things down and creates bottlenecks. If an organization's sole super admin goes on vacation and nobody else has any type of administrative privileges, nothing can get done.

---

68 Grimes, Roger A. "Too many admins spoil your security." InfoWorld, http://www.infoworld.com/article/2614271/security/too-many-admins-spoil-your-security.html. Accessed 15 June 2017.

69 Ibid.

70 Crouse, Michael. "A growing threat: Privileged user abuse." SC Media, https://www.scmagazine.com/a-growing-threat-privileged-user-abuse/article/543080/. Accessed 23 June 2017.

Challenge #2: Implementing the Least Privilege Model in SaaS Environments

Assigning the appropriate level of privileges (either within one SaaS application or across multiple applications) can be a thorny process.

The reason is twofold. First, IT can't view its team's admin privileges across applications in one place—there's no unified view. This is SaaS sprawl at work. Not only are permissions siloed within each individual application, but there is also little to no consistency across applications in either terminology or levels of privileges offered. Each application has unique roles, and it's cumbersome to track the functionality in each role type (for instance, here is a just a sampling of the panoply of roles available in several SaaS applications: "team admin," "user management admin," "support admin," "owner," "admin, "administrator," and "account owner." It's impossible to correctly distinguish all the roles from each other). Additionally, IT has no control over these roles if SaaS vendors add permissions or modify them somehow.

Therefore, employees who only need a few rights may be made into **unwitting super admins**. IT teams may not necessarily want to grant an employee *carte blanche*, but they are left with no other option.

Second, many of the less mature SaaS applications offer only binary options: super admin or end user, with nothing in between—it's all or nothing. Therefore, employees who only need a few rights may be made into unwitting super admins. IT teams may not necessarily want to grant an employee *carte blanche*, but they are left with no other option. IT must either hand over all the "keys to the kingdom" or risk impeding employee productivity, because end users must then wait for IT to make user lifecycle and data management changes.

"The smaller SaaS applications we use have no granularity when it comes to admin permissions. As a result, I'm left in a tough spot," says Tim Burke, Director of IT at BetterCloud. "I have to grant 'super admin' access for billing or other routine admin that in no way needs all the permissions being granted. It's one of those things that used to keep me up at night, but having more granular control of admin permissions puts me, my security team, and our auditors much more at ease."

Here's a common scenario. Employees who are working on a specific task or project will often request admin access or elevated access in general. Full access is given to them but is never taken away, because tracking and auditing admin privileges is difficult to do in native admin consoles. As a result, this leads to an over-assignment of super admin privileges across SaaS applications. Giving employees more rights than needed is ill-advised on many levels, namely because it snowballs into larger security problems down the line. "If you hand out admin privileges like candy, it'll come back to haunt you," writes InfoWorld.[71]

Why? For one, this kind of pattern can have serious ramifications when it comes to security and compliance. "Many organizations underestimate the danger of giving employees full administrative privileges simply to carry out everyday tasks. This access can significantly increase the risk of an attack, as well as the potential for a citation for PCI compliance violations,"[72] writes CIO.com.

"For example, an internal or outsourced IT administrator making unauthorized system changes could unintentionally block antivirus or policy settings designed to protect servers transmitting cardholder data. That's why access to protected data should be granted only when absolutely required. [Least privilege security policies] minimize the risk of a data breach, and help enterprises ensure PCI compliance."[73]

---

71 Grimes, Roger A. "Too many admins spoil your security." InfoWorld, http://www.infoworld.com/article/2614271/security/too-many-admins-spoil-your-security.html. Accessed 15 June 2017.
72 "Leveraging Privileged Identity Management to Support PCI Compliance." CIO, http://www.cio.com/article/3142630/security/leveraging-privileged-identity-management-to-support-pci-compliance.html. Accessed 30 June 2017.
73 Ibid.

Though the least privilege model is a core security principle, actually implementing it in SaaS applications is cumbersome at best and impossible at worst. With SaaS vendors constantly introducing new permissions, it can be difficult to keep up. It's little surprise then that SaaS-Powered Workplaces are 3.5 times more likely than the average workplace to say delegating admin privileges is a challenge.[74] And as SaaS adoption grows, assigning the right level of privileges only becomes exponentially more challenging.

> "[Least privilege security policies] **minimize the risk of a data breach**, and help enterprises ensure PCI compliance."
>
> - CIO.com

Privileged access also becomes a tricky issue in multi-instance SaaS environments. For instance, some admins may only have access to one instance of an application but need access to multiple instances. Conversely, some admins may have access to all instances and should only have access to one. Keeping all of these access needs straight adds an additional layer of complexity to an already complicated environment.

In the wake of Snowden's leaks, the NSA announced it would eliminate 90% of its 1,000 system administrators to reduce the number of people with access to secret information.[75] This case, while not explicitly related to SaaS, is nonetheless a high-profile example of why maintaining a least privilege model is critical. It is a best practice that is instrumental in improving the security posture of all organizations.

74 "2017 State of the SaaS-Powered Workplace Report." BetterCloud, https://www.bettercloud.com/monitor/state-of-the-saas-powered-workplace-report/. Accessed 15 June 2017.

75 Allen, Jonathan. "NSA to cut system administrators by 90 percent to limit data access." Reuters, http://www.reuters.com/article/us-usa-security-nsa-leaks-idUSBRE97801020130809. Accessed 26 June 2017.

Challenge #3: Managing and Monitoring Privileged Accounts

Complicating this problem is the fact that IT often grossly underestimates the number of privileged accounts present in its environment. It is unaware of the magnitude of the sprawl. A 2013 global IT security survey on privileged account security and compliance "found that the number of privileged accounts in an organization is typically 3-4 times the number of employees. When asked to estimate the number of privileged accounts in their organization, 86% of respondents from large enterprises (5,000+ employees) stated they either didn't know how many accounts they had or that they had no more than one per employee. That means at least two out of every three privileged accounts in these organizations are either unknown or unmanaged."[76]

> "... at least two out of every three privileged accounts in these organizations are either **unknown or unmanaged."**
>
> - CyberArk

Respondents in a 2015 global security survey admitted that "the implementation of delegation—the capability to implement a least privilege model of admin activity in which administrators are only given sufficient rights to do their job"—was one of the most critical practices related to critical account management in their organization. But fewer than half said they have a regular cadence of recording, logging, or monitoring administrative or other privileged access.[77]

"To alleviate this risk and ensure these accounts are controlled and secured, it's absolutely crucial for organizations to have a secure, auditable process to protect them," writes CIO.com. "A good privileged account management strategy includes a

76 "CyberArk Survey Shows Majority of Organizations Underestimate Scope of Privileged Account Security Risk." CyberArk, https://www.cyberark.com/press/cyberark-survey-shows-majority-organizations-underestimate-scope-privileged-account-security-risk/. Accessed 23 June 2017.

77 Olavsrud, Thor. "Organizations sloppy about securing privileged accounts." CIO, http://www.cio.com/article/3005613/security/organizations-sloppy-about-securing-privileged-accounts.html. Accessed 23 June 2017.

password safe, as well as least privilege control to protect organizational assets from breaches."[78]

The Delegation Landscape

While delegation vendors exist, none offer the full granularity and customization that SaaS administrators need to have control over their environment. Here is an overview of some of the advantages and disadvantages offered by vendors in the market:

VENDOR CATEGORY	DELEGATION FUNCTIONALITY	ADVANTAGES	DISADVANTAGES
IDaaS	Delegation based on admin roles that the apps natively define and expose. The roles are very coarse and broad	Native definition of admin functionality dictated by each application vendor	Admin capabilities exposed by applications have too much privilege associated with them, which does not help with enforcing least privilege security models. IT teams are left with no choice but to grant all team individuals with super admin privileges
SaaS Application Management and Security Platform	Delegation is fine grained and can be custom applied at the individual data object level for every application	Least privilege models can be applied by granting IT teams exactly the right level of privilege they require. Super admin access to applications can be locked down	Ability to only apply this to the most common IT-sanctioned apps, given there will not be support for "long tail" apps

78 Ibid.

The Solution: Granular Delegation (Role-Based Privileges)

The only way to ensure that employees are granted the right amount of access is to have the ability to implement a least privilege model in SaaS environments.

The permissioning needed at this level relies upon standard CRUD (create, read, update, delete) functions in each SaaS application and data object. But ideally, IT should be able to use a "tabula rasa" approach: start with a blank slate, and then create custom roles with specific, granular privileges for the data objects and controls across SaaS applications. IT can then delegate limited privileges to different business units and employees. This ensures that no employee has excessive access to applications or data.

Data objects (and even specific SaaS applications) have varying levels of sensitivity and criticality. Therefore, having fine-grained delegation capabilities is important for maintaining operational security.

For example, IT might want to delegate read-only access on specific objects in one SaaS application in addition to full CRUD in another SaaS application. This granularity allows employees to subscribe to the least privilege model. Prescribing and enforcing least privilege access is a best practice that is crucial in minimizing security risks in SaaS environments.

For delegation to be possible, all the elements of the framework must be in place. The foundational element Centralization exposes access controls and permissions across various SaaS providers when data objects are ingested and normalized. Once this is done, Discoverability comes into play. IT can then filter information, limiting the specific objects, data types, and applications employees can take action on. This essentially gives IT an access control service, enabling IT to turn specific actions on or off for select users. Other teams (like HR) can then make their own user lifecycle changes without having to rely on IT or risk compromising security.

Use Cases: How Delegation Enables SaaS Management Success

Here are several examples of Delegation use cases:

- The HR team cannot reset passwords, but can create, read, and update users, groups, and org units in one SaaS application (or across multiple SaaS applications) as they onboard new employees.
- The security team can only read, update, and delete files, sites, calendars, and tickets.
- The helpdesk is only given the ability to change group settings.
- The office manager and administrative assistant have edit permissions on a calendar resource (e.g., a conference room), but do not have permission to edit calendar settings (e.g., see only free/busy, make changes to events, etc.).
- Nobody in the company can install any third-party applications except the IT department.
- The helpdesk can change permissions on any files except anything owned by HR or containing the word "Confidential."
- The helpdesk only has admin privileges from 9 AM until 5 PM.
- Users can request a custom access role that IT can also add an expiration date to.

Here are some use cases that combine Delegation with the powers of Automation:

- When HR adds a new employee to an HRIS, the new employee is automatically granted access to the appropriate groups, files, folders, and applications, and the ticketing application automatically generates a checklist ticket, reminding IT of their offline to-do tasks (e.g., prepare a new laptop and keycard for the new employee).

- IT automatically delegates select admin privileges based on a user's department, manager, or title in bulk.

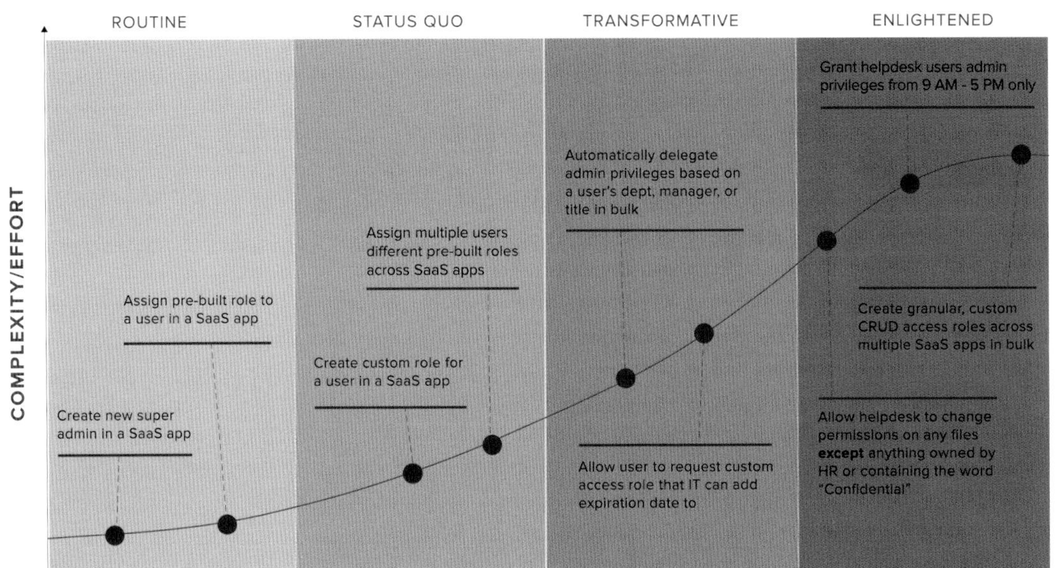

Figure 15

Use Cases: From Routine to Enlightened

Delegation tasks that fall into the "Routine" category include creating new super admins and assigning pre-built roles to a user in a SaaS application. These tasks are fairly easy and can be done in a few steps in a native admin console.

Assigning pre-built roles in bulk across multiple applications, however, becomes a more challenging process. Automatically delegating admin privileges based on a user's department, manager, or title is impossible to do natively but would very useful for IT as it would eliminate a large amount of manual work and enforce the principle of least privilege.

But it is the delegation tasks in the "Enlightened" stage that would bring the most value to IT. Having the ability to scope access roles by time, for example, is not something IT can easily do. For example, IT could grant helpdesk users admin privileges from 9 AM until 5 PM only. This adds a whole new layer to the concept of least privilege and gives IT the additional control it desires over its environment. Another example in the "Enlightened" stage is blacklisting files for users—e.g., allowing helpdesk users to change permissions on any files except those owned by HR or containing the word "Confidential." This type of granular delegation would protect sensitive data while also giving users sufficient rights to do their jobs. ▲

AUDITABILITY

Challenge: Thorough Auditing Capabilities in SaaS Environments

If an organization undergoes a compliance audit, it must produce an audit trail. IT must be able to monitor and track all admin actions in each application to ensure compliance.

However, auditing is a major blind spot for IT. When multiple admins are working in multiple admin consoles, it is very difficult to ascertain who accessed what and when, which actions were taken, which issues were remediated, and how they were remediated. This issue becomes exacerbated when administrators share accounts, which is more common than one might think. A global security survey found that 37% of respondents said multiple admins share a common set of credentials. This bad habit obfuscates who the actor truly is. In the same survey, 31% of respondents said they were unable to consistently identify individuals responsible for administrator activities.[79]

79 Ibid.

The lack of auditability was a major contributing factor to Snowden's success in stealing confidential data from the NSA server. "As a system administrator, Snowden was allowed to look at any file he wanted, and his actions were largely unaudited. He was also able to access NSAnet, the agency's intranet, without leaving any signature. He was essentially a 'ghost user,' making it difficult to trace when he signed on or what files he accessed."[80] Again, while this is not a SaaS-specific case, it still serves as a reminder that auditability is a universally critical element in keeping data safe.

Automation engines and scripts are often not recorded in audit logs. If bots or third-party applications are used in a script, **there is no way to see who ran them or what actions were taken.**

There is no easy way to thoroughly track an admin's actions in SaaS applications. An IT administrator would have to download logs from each SaaS application and parse through them one by one, manually correlating events across all of them. Plus, automation engines and scripts are often not recorded in audit logs. If bots or third-party applications are used in a script, there is no way to see who ran them or what actions were taken.

80 Esposito, Richard and Matthew Cole. "How Snowden did it." NBC News; http://www.nbcnews.com/news/other/how-snowden-did-it-f8C11003160. Accessed 26 June 2017.

The Auditability Landscape

There are many vendors in the market that provide audit logs. However, while they provide copious information about events, they lack sufficient context for IT to extract any meaningful insight. They are also not actionable. IT cannot remediate any issue directly from inside these applications. Here are some advantages and disadvantages:

VENDOR CATEGORY	AUDITABILITY FUNCTIONALITY	ADVANTAGES	DISADVANTAGES
CASBs	API-mode CASBs can provide audit capabilities by aggregating audit logs from disparate SaaS applications	Can support forensics efforts to review user activities for a holistic understanding of actions performed inside SaaS applications	Laborious to sift through audit logs and filter relevant information; often seen as overkill for the problem
SaaS Application Management and Security Platform	Such platforms provide audit logging for all admin actions taken from the platform across disparate SaaS applications	This is a holistic, more pragmatic, and less burdensome approach to centralizing all admin access and control before easily auditing all activity from the same place	Does not capture or aggregate native audit logs. All admin access and operations in applications must go through this platform, and individual SaaS app admin consoles should be minimally used

The Solution: Auditability

To ensure that their organization is in compliance with industry and federal regulations, IT teams needs a way to oversee and audit what every user is doing against all the applications they have access to.

Sarbanes-Oxley (SOX) audits are an example of where logs are crucial. IT departments need to produce reports and documentation for auditors that prove, for example, when specific access privileges were granted or revoked. "Accounting for access (particularly administrative access) to critical systems is an important aspect of SOX compliance. Systems must be configured to capture both administrative and user access, to store the logs for later review and to protect the logs from unauthorized access," writes Varonis.[81]

Auditability relies on the foundational elements of the framework to work. Once the data is centralized into one place, it becomes searchable. This means IT can easily search audit logs (by multiple attributes such as user, activity, event name, date, and more) for a single application or even multiple applications. That information can be filtered and presented in real time. Therefore, IT can maintain one consolidated audit log.

"Compliance auditors will generally ask CIOs, CTOs, and IT administrators [questions like] what users were added and when, who has left the company, whether user IDs were revoked, and which IT administrators have access to critical systems," writes TechTarget.[82]

As such, critical events that should be logged include login attempts (both failures and successes), data accessed, information security events, use of privileges (including advanced privileges), and administrative configuration changes. For this to be possible, there needs to be code inside SaaS applications that necessitates the inclusion of fields vital for auditing purposes. Examples include the actor (e.g., the username, UUID, or API token name of the account taking the action), location, time, date, action, action type, event name, event description, object (the resource that was changed), who it affected, etc. In order to write to the API and commit them as a log entry, these fields must be automatically populated and included. This way, IT will always have thorough, relevant data recorded in their audit logs.

---

81 "SOX: Understanding Sarbanes-Oxley: How to bring your network and data into compliance with the Sarbanes Oxley Act of 2002." Varonis, https://www.varonis.com/learn/sox-sarbanes-oxley-act-compliance-requirements-for-it/. Accessed 26 June 2017.

82 Rouse, Margaret. "Compliance audit." TechTarget, http://searchcompliance.techtarget.com/definition/compliance-audit. Accessed 30 June 2017.

Use Cases: How Auditability Enables SaaS Management Success

Here are some examples of Auditability use cases:

- A disgruntled employee has been grumbling about his job during the past week to the point that HR is concerned about his behavior. IT doesn't know what suspicious actions he has taken (if any), but it knows the general time window and the user in question. They search logs by date and users and see that this employee has mass downloaded data from Dropbox, Salesforce, and Google Drive within three days; forwarded emails to personal accounts; and shared multiple files with a competitor.
- An audit reveals that multiple group administrators have not made any changes to their groups in the past six months. This tells IT that these users likely do not need to be administrators, and thus it revokes their administrator access. This helps implement the principle of least privilege.
- IT runs an audit to get a listing of all its Google+ users. It notices unusual activity: one employee in Malaysia is posting significantly more often than any other employee. IT investigates and discovers that this employee is using her corporate Google+ account to share and propagate ISIS and other terrorist propaganda. She is later arrested and this leads to the arrest of several other ISIS-related individuals **(this is a true story).**
- A contractor's three-month access has just expired. Once their accounts have been deleted, IT automatically receives audit logs containing a detailed activity history to review.
- When passwords are reset in a SaaS application, audit logs are automatically sent via Slack to the security team.

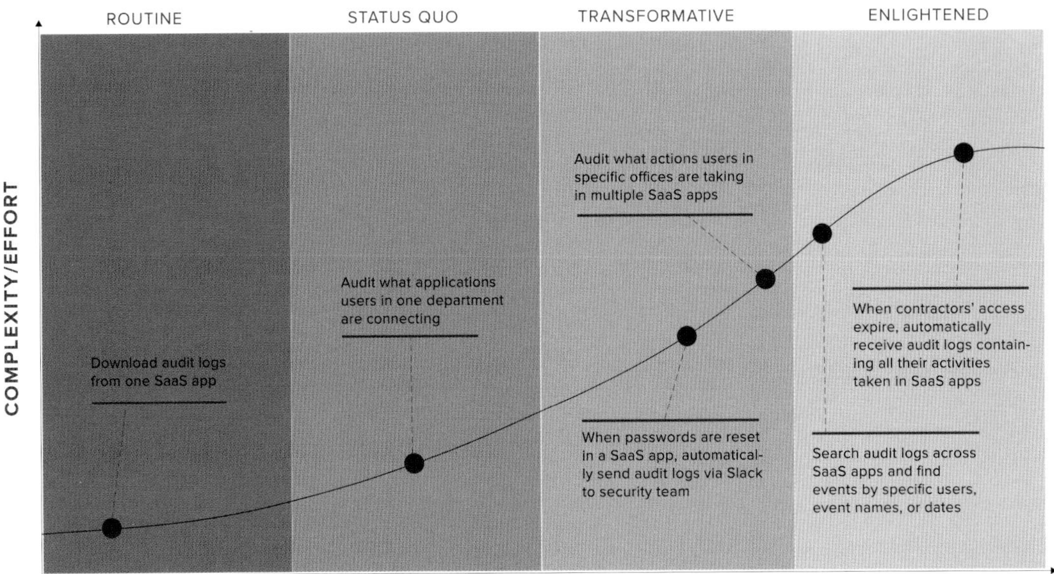

Figure 16

Use Cases: From Routine to Enlightened

IT can download audit logs from one SaaS application quite easily, but these are not tremendously valuable as these logs often lack context. Furthermore, they are not actionable.

Auditing what applications users are connecting to becomes a bit more challenging and time-consuming, but these yield more valuable insights.

Finally, searching audit logs across SaaS applications by event names, users, and dates is flat out impossible to do in native admin consoles because audit logs are not natively searchable or actionable. But this capability would be immensely valuable for IT. Finding events and then taking action directly from the audit log (rather than logging into multiple admin consoles) to remediate a problem would save huge amounts of time. ▲

HOW ALL THE ELEMENTS OF THE FRAMEWORK WORK TOGETHER

Below are various categories of use cases, such as onboarding, offboarding, security, etc. viewed through the lens of this framework. Each example illustrates how each element builds upon the previous one. As the framework progresses, each step becomes increasingly more powerful. These examples below also illustrate what IT can ultimately achieve when the framework is fully executed.

Onboarding Example:

CENTRALIZATION	**View** all new users in an organization.
DISCOVERABILITY	**Identify** all new users created in the past seven days.
INSIGHTS	**Alert IT** when a user is newly created in an HRIS.
ACTION	**Grant new user access** to the appropriate files, folders, calendars, and groups in the appropriate SaaS applications and **send reminder message** to manager to schedule a welcome 1:1 meeting.
AUTOMATION	**Automatically grant new user access** to the appropriate files, folders, calendars, and groups in the appropriate SaaS applications **based on his title, manager, or team**, and **automatically send reminder message** to manager to schedule a welcome 1:1 meeting.

Offboarding Example:

CENTRALIZATION	**View** all users in an organization.
DISCOVERABILITY	**Identify** users who are in the "To Be Offboarded" OU.
INSIGHTS	**Alert IT** when a user is deleted/suspended in an HRIS.
ACTION	**Hide user in directory**; reset user's password; add an auto-reply message; delegate email access to another user; transfer ownership of sites, calendars, files, and groups to another user; revoke access to third-party applications; and suspend/delete user.
AUTOMATION	**Automatically hide user in directory**; reset user's password; add an auto-reply message; delegate email access to another user; transfer ownership of sites, calendars, files, and groups to another user; revoke access to third-party applications; and suspend/delete user in SaaS applications.

Security Example:

CENTRALIZATION	**View** all users who work in the sales department.
DISCOVERABILITY	**Identify** user in the sales department who has downloaded more than 100 records from Salesforce within seven days, is sharing documents with competitors, and is logging in from home on the weekends.
INSIGHTS	**Alert IT** when a user in the sales department has downloaded more than 100 records from Salesforce within seven days, is sharing documents with competitors, and is logging in from home on the weekends.
ACTION	**Suspend account,** send Slack message to supervisor, and open a ticket with the security team.
AUTOMATION	**Automatically suspend account,** send Slack message to supervisor, and open a ticket with the security team.

External User Management Example:

CENTRALIZATION	**View** all external users who have access to organization's data.
DISCOVERABILITY	**Identify** all external users who have @xyz.com in their email address and have access to files with the word "Confidential" in their content.
INSIGHTS	**Alert IT** when external users with @xyz.com in their email address have been part of the organization for more than 30 days.
ACTION	**Remove access** for any external users with @xyz.com in their email address after 30 days of inactivity across SaaS applications.
AUTOMATION	**Automatically remove access in bulk** for any external users with @xyz.com in their email address after 30 days of inactivity across SaaS applications.
DELEGATION	**Create custom roles** for external users with read/update privileges for files and calendars only for 15 days.
AUDITABILITY	**Search for all actions taken by a specific external user** over the past 90 days.

Compliance Example:

CENTRALIZATION	**View** all executives in an organization.
DISCOVERABILITY	**Identify** executives who are forwarding emails to personal accounts (Gmail, Yahoo, Outlook) and sharing documents with external domains.
INSIGHTS	**Alert IT** when executives are forwarding emails to external accounts (Gmail, Yahoo, Outlook) and sharing documents with external domains.
ACTION	**Reset password** for these accounts and **un-share documents with external domains.**
AUTOMATION	**Automatically reset password in bulk** for these accounts and **un-share documents with external domains in bulk**.

License Management Example:

CENTRALIZATION	**View** all users using multiple applications in an organization.
DISCOVERABILITY	**Identify** inactive users who haven't logged into multiple applications in 30 days.
INSIGHTS	**Alert IT** when a user hasn't logged into multiple applications in 30 days.
ACTION	**Delete dormant user's accounts across all SaaS applications** and recycle licenses for new users.
AUTOMATION	**Automatically delete dormant user's accounts across all SaaS applications** and recycle licenses for new users.

Group Management Example:

CENTRALIZATION	**View** all users in an organization.
DISCOVERABILITY	**Identify** users who have changed office location, country, business unit, manager, subsidiary company, and/or employee status (e.g., full-time vs. contractor) in the past seven days.
INSIGHTS	**Alert IT** when a user changes office location, country, business unit, manager, subsidiary company, and/or employee status.
ACTION	When a user changes office location, country, business unit, manager, subsidiary company, and/or employee status, **move them to (and remove them from)** the appropriate groups.
AUTOMATION	When a user changes office location, country, business unit, manager, subsidiary company, and/or employee status, **automatically move them to (and remove them from)** the appropriate groups.

NEXT STEPS

As a next step, IT teams should draw upon this framework to assess their own SaaS environments and identify any blind spots. Some important high-level questions to think about are:

Centralization/Discoverability

What information does IT need to have that is impossible (or very difficult) to obtain in native admin consoles?

When IT spends time tracking down and resolving policy violations, what is the cost of productivity loss?

Insights

How much time and energy is spent reviewing (and ignoring) irrelevant alerts per day?

Of those alerts, how many are actually critical?

Where are the biggest operational security risks and non-compliance areas?

Actions

What are the biggest bottlenecks in IT's day-to-day operations?

How could IT increase its operational efficiency?

Automation

Where is there room to automate and do less manual work?

How could that time be better spent? How would those projects move the business forward?

Delegation and Auditability

How are administrative privileges across SaaS applications currently tracked and controlled?

If IT were to undergo a compliance audit tomorrow, could all the necessary information be pulled?

IT should think critically about how it can apply this framework to close the gap between strategy and execution. IT can create a strategy for management and operational security, but actually making it work is an entirely different story. This framework identifies exactly the elements IT needs to execute on its strategy.

CONCLUSION

SaaS is now a common system of record for organizations. It's revolutionizing the modern workplace. For IT, this shift has far-reaching implications that should be addressed sooner rather than later.

SaaS sprawl is undermining IT's control. Too many IT professionals underestimate the blindness created by SaaS sprawl and the dangers created by excessive privileged access. Ultimately, the way IT manages and operates its SaaS applications needs to change—the stakes are too high for it not to. SaaS data continues to multiply at dizzying speeds and operational security risks continue to increase. It's not a matter of "if," but "when" IT will reach a breaking point.

IT needs a way to reliably manage, secure, and support its SaaS environment. To do so, it must understand exactly how SaaS applications operate and interact with each other. Next, it needs the six elements in this framework—Centralization, Discoverability, Insights, Action, Automation, and Delegation/Auditability—to fill the gaps left behind by native admin consoles. Collectively, these elements empower IT, allowing it to wrangle SaaS data, automate manual work and compliance, avoid alert fatigue, and minimize security risk. By implementing this framework, IT can secure its data and gain an unprecedented level of control and clarity over its SaaS environment.

ABOUT BETTERCLOUD

BetterCloud

BetterCloud is the first SaaS Application Management and Security Platform

BetterCloud enables IT to centralize, orchestrate, and operationalize day-to-day administration and control across SaaS applications. Every day, thousands of customers rely on BetterCloud to centralize data and controls, surface operational intelligence, orchestrate complex actions, and delegate custom administrator privileges across SaaS applications.

New to BetterCloud?

Visit **https://www.bettercloud.com/product** to learn more about BetterCloud, the SaaS application management and security platform that was purpose-built to align with this framework. If you would like to work with a product expert to address the needs of your organization, please go to **https://www.bettercloud.com/demo.**

With BetterCloud, IT departments can...

Quickly and easily sift through **hundreds of millions of data objects** (such as files, groups, and users) in their environment in seconds.

Take action in bulk across SaaS applications (including G Suite, Slack, Dropbox, JIRA, Salesforce, Zendesk, Namely, Box, and Azure AD) from a single hub.

Automate complex sequences across applications for user lifecycle management and on- and offboarding, ensuring accuracy and precision of critical operational processes.

Set security policies to automatically flag and/or remediate data exposure risks or violations.

Implement a true least privilege model by delegating and deploying custom administrator roles for security, helpdesk, and/or HR teams not available in native SaaS admin consoles.

Track and monitor administrators' actions in a centralized place, eliminating the need to sift through multiple applications' audit logs.

Save tens of thousands of dollars by centralizing application users and data, and then eliminating inefficiencies such as unused licenses.

Remediate security risks faster by correlating incidents and patterns through advanced alerting out of a centralized activity event stream.

Shave days off training time for new IT hires by standardizing around a common vocabulary and console for managing and securing SaaS applications.

Empower IT with root access to the information stored across SaaS applications.

Existing BetterCloud customers:

If you would like to speak with your customer success manager about making the most of BetterCloud to manage your SaaS applications, please email **success@bettercloud.com.**

BetterIT is the IT industry's leading online community.

Launched in October 2016, this free Slack-based community is home to thousands of forward-thinking IT professionals who helped develop this framework and are implementing it in their organizations today.

BetterIT is an excellent place to share best practices, ask questions, learn from each other, and discover new resources. It's a live, real-time home for any and all IT-related topics. Community members include IT professionals from BuzzFeed, WeWork, Pinterest, Casper, and more.

To subscribe to BetterIT, visit: **https://betterit.cloud.**

www.betterit.cloud

BetterCloud

MONITOR

The BetterCloud Monitor is the online authority for IT information, research, tips, and trends on the shift to SaaS.

Launched in April 2016, the Monitor aims to empower, educate, and celebrate the modern IT professional as they navigate the move to SaaS. To that end, the award-winning media property features interviews with leading IT experts, original research reports, longform content, and an extensive library of 1,000+ SaaS tutorial videos.

The Monitor receives 500,000 unique visitors per month and has a global audience of technology professionals in 232 countries, islands, and territories. Its content has been featured in publications such as the *Wall Street Journal's CIO Journal*, CIO.com, *Forbes*, *Business Insider*, and more.

The Daily Monitor newsletter, an expertly curated digest of tech news, stories, and tips, reaches an audience of 100,000 IT professionals and SaaS enthusiasts.

To subscribe to the Monitor, visit: https://www.bettercloud.com/monitor.

www.bettercloud.com/monitor